The Remote Renaissance

Thriving in the New Workplace Revolution

By

Cassandra Solstice

The Remote Renaissance

Thriving in the New Workplace Revolution

Table of Contents

Introduction

As the world continues to evolve with rapid technological advancements, so too do the ways in which we approach work. The shift to remote work has been nothing short of a paradigm shift, fundamentally altering the landscape of professional life. This book serves as a guide to understanding and adapting to this new era, where work is no longer confined to the traditional office environment. We embark on a journey to explore the nuances of remote work, offering insights and strategies to harness its potential while navigating its challenges.

The concept of remote work isn't entirely new; it's been slowly gaining traction over the years. However, recent global events have accelerated its adoption, making it a central feature of modern working life. This transition has been accompanied by significant societal shifts, impacting not just how we work, but how we live, communicate, and interact with one another. Remote work has forced us to rethink our traditional notions of work-life balance, productivity, and even what it means to be part of a team.

For professionals navigating the turbulent waters of this remote work revolution, the stakes are both exciting and daunting. On one hand, there's the allure of greater flexibility, the elimination of lengthy commutes, and the opportunity to find harmony between personal and professional lives. On the other, there are challenges—distractions at home, feelings of isolation, and the blurring boundaries that once defined our work and personal time. The pages that follow are designed to equip you with the tools and knowledge needed to thrive in this new landscape.

Remote work has also forced companies to reassess their structures and strategies. Organisations are reimagining their cultures, policies, and practices to fit a distributed workforce. For HR leaders and corporate managers, this transition demands a shift in leadership styles, necessitating new approaches to communication, motivation, and team empowerment. It calls for a balance between maintaining productivity and nurturing wellbeing, as well as fostering an inclusive culture that crosses geographical boundaries.

Technology sits at the heart of this transformation, acting as both catalyst and enabler. The proliferation of collaborative tools and software has made it possible for teams to work together seamlessly, regardless of physical location. Yet, this reliance on technology presents its own set of concerns—data security, privacy, and the need for robust IT infrastructure. As we delve into these aspects, we'll uncover how organisations can leverage innovations to create secure and efficient remote work environments.

One of the most profound shifts brought about by remote work is its impact on human connection. The coffee break conversations, impromptu brainstorming sessions, and subtle non-verbal cues—elements once taken for granted in an office setup—are often missing in virtual interactions. This necessitates a rethinking of communication strategies to ensure that the human element isn't lost. We explore ways to build effective communication channels that foster collaboration and harmony among remote teams.

The rise of remote work also brings opportunities for personal growth and autonomy. Professionals can find themselves with more control over their day-to-day schedules, providing space for creative pursuits and skill enhancement. This newfound independence can be liberating, but it requires self-discipline and motivation. We'll discuss techniques to cultivate these qualities, enabling individuals to harness the full potential of remote work.

Throughout this book, we aim to provide a comprehensive look at the multifaceted impacts of remote work. From psychological well-being to economic ramifications, we cover diverse topics to paint a complete picture of this ongoing transformation. Our goal is to inspire and empower you—whether you're leading a team, managing an organisation, or adapting to work from home—to embrace the changes, overcome the obstacles, and seize the opportunities that remote work presents.

In this intricate dance between tradition and innovation, the future of work is being rewritten. We hope to illuminate this path and help you navigate its complexities with confidence and ease. So, let us embark on this exploration together, as we delve into the many dimensions of remote work and what lies ahead in this ever-evolving landscape.

Chapter 1:
The Dawn of the Remote Renaissance

The world of work is undergoing a transformation, heralding what many call the 'Remote Renaissance'. This seismic shift isn't just a response to recent global events; it's an evolution long in the making. Remote work emancipates professionals, offering flexibility and opening doors to unprecedented levels of inclusivity, creativity, and work-life balance. For HR leaders and corporate managers, this dawn represents a new frontier—one characterised by the need for adaptive strategies and innovative tools that bridge both geographies and generations. Yet, it's more than just about managing change; it's about thriving in it. As the sun rises on this era, it's a call to harness potential, mitigate challenges, and embrace a brave new world where the boundaries of 'workplace' are defined by thought rather than walls.

Understanding the Shift

In the quiet confines of our living rooms and the bustling corners of makeshift home offices, a monumental transformation is taking place. The shift to remote work is not merely about changing locations or swapping a suit for pyjamas. It's a profound reimagining of how work fits into the tapestry of our lives, influencing everything from our daily routines to global economic practices.

At its core, understanding this shift means recognising the myriad ways in which it has reshaped our perceptions of productivity and presence. Where once the physical office was a symbol of discipline and output, the ability to work from virtually anywhere has introduced a dynamic, albeit complex, flexibility. This flexibility offers a double-edged sword; it provides the freedom to craft a workday around personal life, while simultaneously blurring the lines between professional and private spheres.

Moreover, remote work has initiated a significant psychological shift. The newfound autonomy can be empowering yet overwhelming. It challenges long-held beliefs about what it means to be a dedicated employee. In the absence of traditional oversight, professionals now often measure success by outcomes rather than hours spent at a desk. This results-oriented mindset demands a higher degree of self-discipline and accountability, fundamentally altering the employer-employee relationship.

As workplaces evolve, so do the expectations placed on individuals by their employers and themselves. Remote work necessitates a culture of trust, where employees are empowered to take initiative and seek creative solutions independently. However, with this trust comes the increasing need for robust communication strategies and clear articulation of goals and responsibilities, ensuring each team member remains aligned and motivated amid physical disconnection.

The shift also prompts reflection on social connectivity within the workforce. The camaraderie that flourished in office hallways and break rooms has had to adapt. Virtual coffee breaks, digital forums, and online social hours have emerged as new vessels for connection, demanding innovative approaches to foster a sense of belonging in a virtual landscape.

From an organisational perspective, adapting to this shift means re-evaluating the frameworks and technologies in place to support a dispersed workforce. Companies are investing in digital infrastructures and team collaboration tools that facilitate seamless interactions, aiming to forge an experience akin to working side-by-side despite geographical divides. This investment is not just technological but cultural, necessitating leaders who prioritise empathy and connection alongside performance.

Importantly, this shift also interrogates the traditional notions of career growth and development. Employees now have unparalleled access to global job markets, yet at the same time, they must proactively engage in continuous learning to remain relevant. Professional development in this era means leveraging online platforms, webinars, and digital mentorships to upscale skills and seize opportunities that hitherto seemed unreachable.

Through this lens of transformation, one can appreciate the shift as both a challenge and an opportunity. It requires a rethinking of how we define success in the workplace—not just in terms of productivity but in achieving a sustainable work-life balance. Professionals who learn to navigate this balance are not only more fulfilled but also more likely to lead their organisations toward innovation.

Nevertheless, this journey is not without its hurdles. Organisations and individuals alike must confront issues of mental well-being and find ways to manage the emotional toll of isolation and the demands of virtual collaboration. This calls for an environment that emphasises psychological safety and support, where discussing challenges openly is as encouraged as celebrating successes.

As we delve deeper into understanding this shift, it becomes evident that adaptability is the cornerstone of thriving in this new era. Embracing change with resilience enables both individuals and businesses to unlock unforeseen potentials. The remote renaissance, then, is not simply a response to immediate necessity but a harbinger of a more dynamic, inclusive future of work.

The Remote Renaissance

In reconciling the old with the new, this seismic shift presents a narrative of hope and possibility—a call to action for all participants in the world of work to rethink, reimagine, and reinvent their approaches. The path forward, illuminated by a greater understanding of remote work dynamics, promises a tapestry of experiences where innovation thrives, and the boundaries of possibility are continuously expanded.

Chapter 2:
Technology as the Catalyst

As the digital age continues to unfurl its myriad possibilities, technology stands as the pivotal force driving the remote work revolution. No longer confined by the physical limitations of traditional office spaces, professionals now find themselves equipped with an arsenal of innovative tools that connect continents and cultures seamlessly. The once-daunting distance between team members is bridged by intuitive software and cutting-edge platforms that not only enhance productivity but also empower individuals to collaborate in ways previously unimaginable. With this technological surge, workers worldwide experience an unprecedented level of flexibility, redefining what it means to work effectively. It is this transformative power of technology that is reshaping industries, fostering creativity, and ultimately, creating a new paradigm for what work can achieve. However, as we navigate these new waters, it's essential to harness these advancements thoughtfully, ensuring they serve to elevate human potential rather than merely automate it.

Innovations Enabling Remote Work

The transformation of remote work from a novelty to a norm has been propelled by groundbreaking innovations that reshape the way we connect, collaborate, and create. From cloud-based platforms that bridge geographical divides in an instant, to AI-driven tools that personalise and streamline workflows, technology has not just caught up with the demands of remote work — it leapfrogs them, providing new possibilities that redefine productivity and engagement. The rise of virtual reality meetings hints at a future where presence no longer requires proximity, while advanced cybersecurity measures ensure our virtual workplaces remain secure sanctuaries. These innovations empower professionals to navigate their roles with confidence and creativity, paving the way for a future where work is defined not by location, but by potential. As we adapt to these changes, we unlock new avenues for growth and connection, demonstrating the limitless potential of human ingenuity in the digital age.

Tools and Software for the Remote Workforce have become the backbone supporting professionals as they transition from traditional office spaces to the expansive, borderless world of remote work. With the rapid advancements in technology, these tools have evolved far beyond simple communication and task management, redefining how we think about productivity, collaboration, and innovation. Whether it's ensuring seamless communication among teammates scattered across time zones or enhancing focus in a home environment filled with distractions, the right tools can make all the difference in maintaining both efficiency and morale.

As we navigate through this digital landscape, the ubiquity of *collaboration platforms* has transformed how teams interact. Platforms like Slack, Microsoft Teams, and Zoom are no longer just tools; they've become virtual meeting rooms and social watercoolers, playing crucial roles in fostering workplace culture and connection. These platforms offer features such as video conferencing, instant messaging, and file sharing all in one convenient interface, making it easier to recreate the spontaneity of in-person interactions. They have forged a path for inclusivity, enabling conversations that might not happen in a standard office setting and allowing quieter voices to be heard.

Task management software has also heralded a new era of getting things done. Tools such as Asana, Trello, and Jira have made it possible to manage workloads with precision, offering users the ability to track projects, assign tasks, and monitor deadlines efficiently. This software helps decentralise oversight, empowering team members to take ownership of their roles and encouraging accountability. By providing visual overviews of projects, they inspire clarity, reduce the clutter of endless emails, and help teams focus on what truly matters. This reshapes the traditional managerial hierarchy, providing a democratic framework where each team member's contributions are visible and valued.

One cannot overlook the critical role that *cloud storage solutions* have played in enabling remote work. Services like Google Drive, Dropbox, and OneDrive have revolutionised access to shared files and documents, allowing teams to work collaboratively in real-time, regardless of their geographical location. These services have not only made physical office spaces less of a necessity but have also secured businesses against the risks associated with local data storage. The ease of access and the peace of mind that comes with knowing that work is securely backed up and readily available wherever you are is empowering for both individuals and organisations.

Security is another area where technology has stepped up to the challenge. As remote work became more prevalent, ensuring that company data remains secure yet accessible has become paramount. VPNs (Virtual Private Networks), two-factor authentication, and advanced encryption protocols have become standard as companies adapt to new security needs. These technologies protect the integrity of sensitive information and foster trust between employees and employers. They assure professionals that their work and personal data are safe from prying eyes, enabling them to engage fully without fear of data breaches or privacy violations.

Of course, none of these tools would be meaningful without the infrastructure to support them. High-speed internet, robust Wi-Fi networks, and reliable hardware are the unsung heroes behind the scenes, ensuring that these software solutions run smoothly and efficiently. These foundational elements, though often taken for granted, have been the lifeline of remote workers, allowing for uninterrupted workflows and high productivity levels. As technology continues to advance, we're seeing more investment in these areas, resulting in faster, more stable connections that further enhance our ability to work from anywhere.

In addition to practical tools that facilitate work, it's equally important to consider the software aimed at enhancing *well-being*. Wellness apps and platforms such as Headspace, Calm, and virtual fitness classes have become integral to maintaining mental and physical health in a remote setting. These tools encourage regular breaks, mindfulness, and physical activity, reminding individuals to prioritise their health amidst their often-demanding work schedules. They create opportunities for balance and provide resources that work alongside our daily routines, adapting to individual needs and preferences.

The collection of these tools represents a robust ecosystem, setting the stage for continuous innovation and adaptation. As more professionals embrace the flexibility and potential of remote work, the development of new and specialised software will likely accelerate. We can expect to see further advancements that will address specific challenges of remote work, such as fostering creativity from a distance and nurturing team spirit without physical proximity. As technology evolves, so does our understanding of human interaction and productivity, pushing us to rethink and redefine what it means to work effectively.

In conclusion, the tools and software shaping the remote workforce are much more than mere convenience. They're reshaping the global work environment, offering individuals the flexibility and freedom to work in ways that best suit them while sustaining organisational goals. They're catalysts, sparking new ideas, creating connections, and ultimately forging a path toward a future where work is not defined by location but by potential and capability. As we continue to embrace remote work, it's inspiring to envision the possibilities that lie ahead, driven by the endless innovations in technology.

Chapter 3:
Redefining Work-Life Balance

As the traditional office setting dissolves into a dynamic, remote-working reality, redefining work-life balance emerges as both a challenge and an opportunity. This digital era, unbound by geographical constraints, tempts us to blur the lines between work and personal life. Yet, it offers a profound chance to shape our routines in ways that resonate with our priorities rather than rigid schedules. In the heart of this transformation lies a call for intentionality and self-awareness, inspiring us to harmonise productivity with personal well-being. By recognising the moments that truly matter, professionals and leaders alike can inspire a paradigm shift, moving from merely surviving work obligations to thriving in a customised blend of professional and personal fulfillment. It's about finding that sweet spot where aspirations meet contentment, and redefining success not by hours logged in, but by lives enriched through meaningful balance.

Challenges and Opportunities in Balancing Act

In the evolving landscape of remote work, achieving a harmonious work-life balance stands out as both a significant challenge and an exciting opportunity for professionals worldwide. With the physical boundaries between home and office slowly dissolving, it becomes imperative for individuals to carve out their own routines that cater to both professional obligations and personal well-being. Flexibility in working hours offers the freedom to attend to personal life while

13

remaining productive, yet it often blurs the lines, leading to overwork and potential burnout. Embracing this new norm requires intentional boundary-setting and self-awareness, enabling professionals to prioritize tasks and manage their time effectively. Simultaneously, while technology facilitates remote work, it also demands a high degree of digital literacy and self-discipline to avoid the pitfalls of constant connectivity. The opportunity lies in the potential for increased job satisfaction and improved mental health when harmony is achieved, making it crucial for organisations and employees alike to invest in strategies that support an adaptable and balanced approach to this modern work dynamic. As we navigate this journey, the success of the balancing act hinges on embracing the fluidity of remote work while maintaining a sharp focus on personal and organisational goals.

Strategies for Maintaining Balance As the boundaries between our professional and personal lives continue to blur, the ability to maintain a harmonious balance becomes not just a skill but a necessity. It's undeniable that the shift to remote work offers both challenges and opportunities. Those who navigate this terrain successfully often find that their strategies are born out of creativity and adaptability. At the core of this balancing act lies the delicate interplay between our daily responsibilities and personal well-being—the cornerstone of productivity and satisfaction in a remote working environment.

One effective strategy is the conscious creation of physical and mental boundaries. Establishing a dedicated work space, even if it's just a specific corner of your living room, can signal a mental shift between home and work life. The importance of rituals cannot be understated. Simple acts like starting the day with a cup of coffee in "work mode" or shutting down your computer at the end of the day are invaluable in maintaining the fine line between personal time and work commitments.

Equally important is the practice of time management. The freedom remote work affords can become a double-edged sword if not wielded wisely. Tools like digital calendars and time-blocking apps are instrumental in maintaining structure. Breaking the day into chunks dedicated to specific tasks ensures that work doesn't bleed into personal time. Additionally, taking advantage of flexible hours to match your peak productivity periods can turn the challenge of self-discipline into an opportunity for enhanced output and less burnout.

Moreover, facilitating effective communication is critical. Technology, while a bridge for remote work, can become an overwhelming flood of notifications and distractions. Clearly setting communication expectations with colleagues—such as designated times for checking emails or responding to messages—can help maintain focus. Tools like project management software or virtual meeting platforms can streamline interactions and reduce unnecessary interruptions, fostering a more balanced workday.

Social connections, though seemingly distant in a remote setup, play a pivotal role in maintaining balance. Virtual coffee breaks or regular informal check-ins with colleagues can recreate the communal aspects of in-person work environments. These interactions not only alleviate feelings of isolation but also nurture professional relationships, which, in turn, can make collaboration more seamless and enjoyable.

Furthermore, incorporating regular breaks and leisure into the workday is not just recommended but essential. The Pomodoro Technique, for instance, encourages intervals of focused work followed by short breaks, enhancing concentration while safeguarding mental health. During these breaks, engaging in physical activity or meditation can refresh the mind and body, providing a counterbalance to the sedentary nature of remote work.

The intentional integration of personal development activities into the weekly schedule can also foster balance. Investing time in learning a new skill or hobby can counteract work-related stress and spark creativity. Whether it's a professional skill related to career development goals or a completely unrelated personal interest, the benefits extend beyond relaxation, often contributing to one's overall sense of fulfilment.

Companies, too, play a role in supporting their employees' quest for balance. Offering flexible scheduling options and fostering a culture that respects personal time can create an environment where employees feel valued and understood. In-person or virtual wellness programmes further underline a company's commitment to employee well-being, offering guidance and resources for maintaining balance in a remote setting.

The road to achieving work-life balance in a remote work context is undoubtedly fraught with challenges. However, these very challenges provide a fertile ground for opportunity. It calls for a willingness to experiment with new ways of working, a courage to set boundaries, and an openness to embrace changes. In doing so, professionals can not only find balance but thrive amid the complexity of modern work dynamics.

Ultimately, mastering the balance in this new era requires a shift in perspective. Viewing work and life as complementary rather than competing entities allows for a more harmonious approach. It's about shaping a personalised strategy that fits one's unique lifestyle, responsibilities, and aspirations. With mindful practice and a supportive network, maintaining balance is not just a dream but an achievable reality.

Chapter 4:
The Psychology of Remote Work

Embracing remote work has shifted the psychological landscape of the professional world in ways we are only beginning to understand. As traditional office walls dissolve, the mental contours of a home workspace delineate new emotional and cognitive experiences. This shift often uncovers profound energy—a mix of liberty and isolation—that demands attention. For professionals, the freedom from commutes and rigid schedules can spark creativity and motivation but may also blur boundaries, enhance stress, or amplify feelings of loneliness. Navigating these complexities requires insight into personal needs and adaptable structures that can foster resilience and mental well-being. Organisations, in turn, have a responsibility to support their remote workforce by cultivating environments—be they virtual or physical—that nurture mental health, encouraging connection, empathy, and productive collaboration. The journey is not without its hurdles, but with intentional strategies, the potential for a fulfilled, balanced digital workplace is well within grasp.

Mental Health Implications

The shift to remote work has brought mental health into sharp focus, making it a pivotal consideration for professionals and organisations alike. While working from home offers unparalleled flexibility, it also blurs the lines between personal and professional spheres, which can lead to stress and feelings of isolation. For some, the absence of physical interaction and routine may amplify anxiety or trigger new mental health challenges. Yet, this transformation also presents an opportunity to foster emotional resilience and adaptability. Embracing mindfulness practices, leveraging digital support networks, and prioritising open communication can create an environment where mental health not only sustains but thrives. It's time professionals and managers champion these initiatives, turning potential pitfalls into pathways for enhanced wellbeing in this evolving remote landscape.

Coping Mechanisms and Support Systems delve into how the mental health implications of remote work necessitate new strategies for resilience and adaptation. As the boundaries between personal and professional life blur, the stressors of remote work emerge from unexpected places. These changes require us to embrace new coping mechanisms and foster support systems that seamlessly fit into our lives, ensuring mental well-being remains a priority.

Foremost among the adjustments is recognising the need for a structured environment even within the perceived freedom of remote work. Establishing a routine can anchor individuals amidst the fluidity of home and work life. It's not merely about sticking to work hours; it's about crafting a rhythm that resonates with personal productivity peaks and troughs. Whether it's a morning ritual or a midday break, creating predictable patterns can provide stability in an otherwise shifting landscape.

However, routines alone aren't a panacea. It's crucial to cultivate a mindset that embraces flexibility. Remote work's nature often means that change is a constant companion, and being adaptable becomes a valuable asset. This adaptability isn't just about managing tasks. It extends to managing expectations, both for oneself and from others. When flexibility and routine work in tandem, they create a resilient framework that can withstand the ebb and flow of remote work's demands.

Equally significant is the need for emotional intelligence in navigating remote work's mental health challenges. The absence of physical cues and face-to-face interactions requires a heightened awareness of one's emotions and those of colleagues. It's about reading between the lines of emails and messages, understanding when to probe deeper or offer support. Developing this sensitivity can prevent misunderstandings, reduce stress, and pave the way for more meaningful connections, despite the physical distance.

The integration of mindfulness practices into daily routines can also serve as a powerful coping mechanism. Mindfulness encourages a deep awareness of the present moment, helping individuals manage stress by focusing on what's within their control. Simple techniques, like deep breathing exercises or brief meditation, can ground the mind and alleviate anxiety, providing a buffer against remote work's potential overwhelm.

One cannot overlook the role of technology in supporting mental health and resilience in remote work settings. Apps designed for mental wellness have proliferated, offering everything from guided meditations to virtual therapy sessions. While technology can amplify stress through constant connectivity, when harnessed wisely, it can also open doors to essential mental health resources. These digital tools provide flexibility and privacy, key components for those seeking help without traditional barriers.

Building robust support systems is another cornerstone of coping in the remote work era. For many, this involves recalibrating how they interact and rely on their professional networks. Regular check-ins with colleagues, whether through informal chats or scheduled meetings, can foster a sense of community and shared purpose. A culture of open dialogue, where employees feel safe expressing their challenges, is vital. This not only creates a support system but also nurtures an environment of trust and understanding, crucial for mental well-being.

Beyond the workplace, personal support networks play an essential role in balancing the mental strains of remote work. Friends, family, and peers outside of work can provide perspectives and emotional support that might be lacking in professional settings. Encouraging a blend of interactions—both professional and personal—ensures that individuals have different avenues to express themselves and seek support.

The importance of physical activity cannot be understated as a coping mechanism for mental health. Remote work often leads to a sedentary lifestyle, which can contribute to mental fatigue. Incorporating regular exercise into daily schedules can boost mood, reduce anxiety, and enhance overall mental resilience. It's not about achieving athletic greatness; it's about prioritising movement as a means to counterbalance the demands of remote work.

As organisations continue to explore ways to support remote employees, the value of mentorship and coaching becomes clear. Connecting employees with mentors can provide guidance, reduce feelings of isolation, and promote personal and professional growth. These relationships can act as a bridge, offering support and fostering resilience, helping employees navigate the unique challenges of remote work.

However, not all coping mechanisms need to be complex or high-tech. The simple act of taking regular breaks can work wonders for mental health. The temptation to overwork is prevalent when home and office converge, but maintaining short, frequent pauses helps sustain energy levels and reduces stress. This simple strategy promotes a healthier work rhythm, encouraging employees to return to tasks with renewed focus and clarity.

Recognising the need for support systems and adaptive coping strategies is half the battle won. The other half lies in prioritising their implementation and continuous evaluation. The mental health challenges of remote work are evolving, and so must our coping mechanisms. By fostering a strong support system and integrating diverse strategies, individuals and organisations can effectively manage the psychological demands of the modern remote workforce.

Chapter 5:
Communication in the
Virtual Workplace

In the ever-expanding landscape of remote work, effective communication stands as the linchpin to professional success. It's not just about relying on technology; it's about weaving a tapestry of clear, meaningful interactions that transcend screens. Embracing virtual communication requires us to be more intentional, picking up on the subtleties that in-person cues once conveyed effortlessly. The secret to thriving in this environment is crafting channels that not only facilitate exchange but also build trust and collaboration among remote teams. By investing in robust digital tools and devising strategies that promote transparency and inclusivity, businesses can transform isolated workdays into seamless collaborative experiences. As we continue to redefine the boundaries of workplace interaction, our ability to adapt and innovate becomes our greatest asset. The journey is complex, but by fostering open lines of communication, we're setting the stage for a future where virtual connections are just as impactful as physical ones.

Building Effective Communication Channels

In a world where virtual work has become the norm, constructing robust communication channels is essential for fostering genuine connection and collaboration. Strong communication isn't just about having the right tools—it's about developing an empathetic culture that encourages transparency and trust. By integrating deliberate strategies to personalise interactions, organisations can create an environment where everyone feels heard and valued. To achieve this, it's crucial to blend asynchronous and synchronous communication intelligently, ensuring clarity and fostering a sense of community. Moreover, encouraging feedback and refining communication practices keeps the channels flowing efficiently. This isn't just about maintaining productivity; it's about nurturing a workspace where ideas can flourish, adapting to individuals' needs, and ultimately crafting a resilient, interconnected workplace ready for future challenges.

Tools to Enhance Interaction are essential in building effective communication channels, especially in today's ever-evolving virtual workplace. The tools at our disposal have revolutionised the way we interact with colleagues, breaking the barriers of time and space. With just a few clicks, individuals from diverse geographies can collaborate as if they are in the same room. These tools aren't just about getting tasks done—they're the heartbeat of company culture, allowing employees to share ideas, provide feedback, and maintain a sense of community. As we integrate these tools into our work lives, they become indispensable allies in achieving professional goals and personal well-being.

A successful virtual workplace thrives on reliability and accessibility, making the choice of communication tools a top priority for organisations. Platforms like Slack, Microsoft Teams, and Zoom have become titans in this field, each offering unique functionalities to cater to various organisational needs. They provide seamless interfaces for instant messaging, video conferencing, and project management, enabling teams to stay connected regardless of location. Importantly, these tools help maintain clarity in communication, reducing the risk of misunderstandings that can occur in email exchanges.

Beyond functionality, these tools also embrace the importance of context and nuance in virtual communication. Features such as emojis, GIFs, and video calls add a personal touch to digital interactions, helping convey emotion and maintain team morale. For instance, a team member celebrating a personal milestone can share their excitement visually, keeping the human element alive in the digital space. This emotional connectivity not only strengthens professional bonds but also nurtures a more engaging and inclusive work environment.

However, it is not enough for tools to merely exist; adoption and integration are crucial. Ensuring that all employees, regardless of their technical prowess, feel comfortable using these tools will maximise their effectiveness. Offering training sessions and user-friendly guides reduces resistance and encourages exploration of these platforms' capabilities. Organisations can also foster an ongoing feedback loop, encouraging users to share their experiences and suggest improvements, thereby continuously refining their communication framework.

Security is another pillar in choosing the right tools for interaction. The increase in remote work has heightened concerns about data breaches and unauthorised access. Hence, platforms with robust security protocols, including end-to-end encryption and multi-factor authentication, are non-negotiable. Protecting sensitive company information while maintaining user-friendly access is a delicate balance that, when achieved, can assure both management and staff that their interactions are secure.

Besides fostering direct communication, interaction tools can be pivotal in project collaboration and management. Software like Asana and Trello, integrated with communication platforms, allow employees to track tasks, deadlines, and progress in real time. These tools offer transparency and accountability, reducing the potential for operational inefficiencies. Moreover, they can be customised to fit different project requirements, ensuring they support rather than dictate workflow.

In considering the integration of these tools, it's essential to factor in comprehensive inclusivity. This encompasses ensuring that the chosen communication platforms are accessible for employees with disabilities. Features like screen readers and closed captions in video content provide an equal opportunity for all team members to participate actively. Such inclusivity not only adheres to ethical standards but also enriches the workplace by drawing from a more diverse pool of ideas and perspectives.

The evolution of tools that enhance interaction doesn't stop at existing platforms. Emerging technologies and innovations continue to redefine how we think about virtual communication and collaboration. Artificial intelligence, for example, is increasingly being used to automate mundane tasks and provide real-time insights, allowing employees to focus more on creative and strategic aspects of their work. Meanwhile, virtual and augmented reality offer immersive experiences that can simulate physical presence in a digital environment, opening new possibilities for training and brainstorming sessions.

Yet amidst these technological advancements, it is important to not lose sight of the human element. Tools should serve to enhance human interaction, not replace it. The most successful implementations are those that integrate technology with empathy, where organisations prioritise the well-being and emotional health of their employees. Innovative tools can foster a sense of community and belonging, essential components in driving employee satisfaction and retention in the virtual workspace.

As we look to the future, the potential for digital tools to transform communication in the virtual workplace is boundless.

Leaders must remain agile, open to evolving technologies and willing to invest in training and support. Employees, in turn, should be encouraged to embrace these tools, recognising them as opportunities for growth and connection. Together, they can cultivate a dynamic ecosystem where communication tools are not just mechanisms of dialogue, but catalysts for innovation and harmony in the virtual realm.

Chapter 6:
Leadership in Remote Teams

In the realm of remote work, leadership is evolving from traditional command structures to more adaptive, flexible approaches. Leaders must cultivate a sense of trust and autonomy within their dispersed teams, using technology as a bridge to maintain connection and engagement. Remote leadership calls for a balance between guiding and empowering, where leaders inspire motivation through visibility and relatability rather than physical presence. It's about transforming challenges into opportunities to innovate and motivate, ensuring that each team member feels valued and part of the organisational mission. In this setting, adaptability is paramount; the ability to listen intently and tailor leadership styles to meet diverse team needs defines the new-age leader. By embracing a collaborative mindset, leaders can foster an environment where creative thought flourishes, and individual contributions amplify collective successes. This shift represents not just a change in where we work but fundamentally how we lead, compelling leaders to rethink their strategies to inspire and drive performance across geographies.

Adapting Leadership Styles

In the dynamic world of remote teams, adapting leadership styles isn't just an option—it's a necessity. Great leaders realise that the shift from in-person to virtual environments demands flexibility and innovation in guiding their teams. Remote leadership requires a blend of empathy, clear communication, and trust-building to foster a sense of belonging and engagement from afar. It's about empowering team members while creating a supportive culture that values their autonomy. Leaders must embrace technology to deliver inclusive and dynamic experiences, ensuring that every voice is heard and valued. By shifting focus from traditional hands-on management to a more transformational approach, leaders not only navigate the challenges of remote work but also seize the opportunity to inspire creativity and resilience within their teams.

Empowering and Motivating Teams Leadership in remote teams requires an agile approach, sensitive to the distinct landscapes of virtual environments where direct oversight and in-person interactions are limited. Yet, this doesn't mean inspiration and empowerment should be any less potent. In adapting leadership styles, it's crucial to cultivate a sense of autonomy among team members while ensuring each person feels integral to the collective mission. This section delves into how leaders can intricately balance guidance with freedom to empower teams consistently, fostering a motivated, cohesive, and innovative workforce despite the distances.

Remote leaders must first grasp the nuances of empowerment. It starts with trust, a fundamental building block often taken for granted in physical office environments. In the virtual world, trust isn't merely a sentiment; it's a strategy. Teams thrive when they feel valued and trusted to self-direct, reducing the micromanagement impulses that can stifle creativity and morale. Leaders who trust their teams allow individuals to explore new ways of working, ultimately leading to novel problem-solving approaches and heightened engagement.

Another vital aspect is effective communication, tailored to remote settings. Clarity and openness in interactions can reinforce empowerment by ensuring every team member comprehends their role and the overall vision. This, coupled with regular feedback loops, enables employees to feel heard and appreciated, playing a key role in maintaining high levels of motivation. Notably, feedback in remote teams should be constructive and delivered in a way that builds rather than diminishes individual confidence.

Empowerment also stems from the right resources and tools. Providing team members with access to the latest technology and training not only enhances productivity but also signals an investment in their personal and professional growth. Leaders must seek avenues to remove barriers that inhibit team performance, thus equipping individuals with the capabilities to take initiatives and break new ground within their spheres.

Additionally, leaders need to incorporate flexibility into their management tactics. The remote setting inherently reduces traditional boundaries between work and life, necessitating a need for adaptable work schedules that accommodate diverse lifestyles. Flexibility increases job satisfaction, allowing employees to perform at their best when they can balance their responsibilities seamlessly. This empowerment to control their schedules often results in increased loyalty and enthusiasm for work.

Despite the reliance on digital communication, creating an emotional connection remains indispensable. Leaders should strive to build a supportive community where team members feel part of something greater. This can be achieved by organising virtual team-building activities or informal chats where participants can share personal stories, fostering an environment of empathy and understanding. Such connections help in tightening the bonds of teamwork, even when physical distances are vast.

A significant element of motivating remote teams is recognising and celebrating achievements, no matter how small. Frequent recognition helps sustain motivation, fostering a culture where accomplishments are highlighted and everyone is inspired to excel.

Encouraging peer recognition can be equally powerful, as it creates a sense of shared success and appreciation within the team.

Opportunities for professional development should not be overlooked. Leaders who invest in their team's growth are likely to motivate and engage them long term. By offering avenues for skill enhancement, such as online courses or workshops, leaders can stimulate a continuous learning culture that not only satisfies the individual's ambitions but also aligns with organisational goals.

Finally, embodying and demonstrating resilience as a leader can inspire teams to persist through challenges. Remote work settings can amplify the impacts of unforeseen difficulties. Leaders who model tenacity offer their teams a framework for overcoming obstacles together. This resilience, coupled with a strategic approach to problem-solving, empowers teams to navigate and innovate through complex situations.

Ultimately, the art of empowering and motivating teams in a remote environment resides in adapting one's leadership style to be inclusive, adaptive, and empathetic. By placing emphasis on creating trust, ensuring effective communication, providing necessary tools, allowing flexibility, fostering emotional connections, recognising achievements, encouraging professional growth, and demonstrating resilience, leaders can profoundly impact their team's engagement and performance. As organisations continue to embrace the evolving dimensions of remote work, the role of inspiring and empowering leadership becomes ever more critical in charting paths to success.

Chapter 7:
Building Culture from Afar

As remote work transforms from novelty to norm, the quest to cultivate a thriving corporate culture at a distance gains paramount importance. In this new era, the essence of a company's ethos isn't confined within the walls of a traditional office but rather woven through digital threads that connect a dispersed workforce. Creating such culture demands intention and innovation; it requires leaders to leverage virtual tools to build connections, promote shared values, and foster a sense of belonging. This isn't solely about Zoom calls or Slack channels, though they play a part. It's about crafting rituals, celebrating wins, and recognising milestones in ways that resonate, even if a handshake isn't possible. When done right, remote culture becomes a magnet that attracts talent, retains employees, and ignites inspiration across any distance imaginable. The key lies in understanding that culture is not a place—it's a practice, an ongoing narrative crafted by every member of the organisation, one interaction at a time.

Cultivating Corporate Culture

In the realm of remote work, cultivating a strong corporate culture isn't just a nice-to-have; it's essential for thriving in an often fragmented environment. Envision a workplace where shared values and norms transcend physical boundaries, weaving a tapestry of connection and common purpose. It's about creating a sense of belonging, even when we're miles apart. This involves crafting narratives that resonate, aligning individuals with the company's mission, and nurturing trust through authenticity and transparency. The cultural touchstones of an organisation are its glue, ensuring every team member feels seen, valued, and motivated. Leaders and employees alike must champion these efforts through intentional actions and open dialogue, fostering an environment where innovation and collaboration flourish. By embedding this culture into the everyday virtual interactions, organisations can not only maintain but enhance their ethos, making distance merely a detail in an otherwise cohesive corporate journey.

Activities and Engagement Strategies, a vital subsection in the journey of cultivating corporate culture from afar, addresses the critical need for intentionality in building a vibrant workplace culture that transcends physical boundaries. In the remote work arena, conventional office interactions—those impromptu hallway chats, collaborative brainstorming sessions, and after-work social gatherings—have all but vanished. Yet, the essence of camaraderie and shared purpose remains crucial. How, then, can organisations foster these bonds in a world where digital screens replace physical proximity? This question is not just vital; it's transformative.

To start, let's delve into the significance of shared experiences, even those that occur virtually. Activities designed to engage remote employees must be both inclusive and meaningful. This is no small feat, given the diversity in backgrounds and time zones. A well-structured framework of activities can help bridge these gaps. Virtual team-building exercises, for example, can integrate a mix of work-related and social interaction, crafting an atmosphere where individuals feel valued not just for their output, but for their presence.

An effective strategy here involves leveraging technology to create interactive and immersive experiences. Imagine the potential of virtual reality (VR) meetings where team members appear as avatars in a digital conference room, or online platforms that allow creative expression through collaborative projects or hackathons. The technology exists, and when used effectively, it can foster a sense of belonging.

Moreover, structured engagement activities like virtual coffee breaks or themed chat channels serve as informal arenas for employees to connect on a personal level. Such activities encourage open dialogue and help dismantle the barriers of formality often associated with digital communication. Here, leaders and managers play a pivotal role. Their participation in these informal events sends a powerful message that engagement is not merely a 'nice-to-have', but a core component of the company's ethos.

Scheduled celebrations of successes, whether big or small, can also work wonders in solidifying corporate culture. Regularly acknowledging achievements and milestones through virtual awards or spotlight sessions ensures that employees' hard work does not go unnoticed. It's about building a tradition, a company-wide habit of celebration that boosts morale and reinforces a culture of recognition.

Furthermore, creating an environment that promotes continuous learning and development is inherently engaging. Organisations can establish dedicated 'learning days' where employees are encouraged to step away from their everyday tasks and immerse themselves in knowledge enhancement. By fostering a culture that values personal and professional growth, companies strengthen their bond with employees, making them feel as though they are part of something larger, something evolving.

Another engagement strategy is the innovation hub model, a digital space where ideas flow freely, leading to potential breakthroughs. Encouraging employees to contribute to innovation labs or forums that emphasise creativity can foster a sense of ownership and investment in the company's future. This can be coupled with a feedback loop, where leaders actively seek insights from employees to shape new initiatives, thereby creating a collaborative cycle of growth and development.

Engagement is also about adaptability, particularly in recognising cultural and geographical diversity within teams. Organisations should provide platforms for employees to share their cultures and traditions, perhaps through virtual cultural exchange days or cooking classes. This not only enriches the cultural tapestry of the organisation but also builds empathy, understanding, and unity.

Wellness initiatives play a crucial role in engaging a remote workforce, particularly in an era where work-life boundaries blur easily. Providing access to virtual wellness activities, such as yoga classes, mindfulness sessions, or even book clubs, can contribute to a holistic engagement strategy. By recognising the importance of mental and physical well-being, organisations demonstrate care for their employees' overall health.

Ultimately, the effectiveness of these activities hinges on leadership's commitment to nurturing an engaging virtual environment. Leadership should actively solicit feedback, continuously seeking to refine and enhance these programs based on employee experiences. Empowerment and autonomy, emboldened by an inclusive activity framework, are keys to transforming engagement strategies from mere activities to potent cultural catalysts.

To conclude, while the challenges of building culture from afar are ample, they are not insurmountable. Energy and creativity in designing activities and engagement strategies will ensure remote work does not dilute corporate culture but reinforces it. The digital realm, despite its lack of physical presence, has the potential to create connections as strong, if not stronger, than their physical counterparts when approached with intention and innovation. As remote work continues to define the future, so too must our efforts to cultivate a culture that nurtures and inspires from any distance.

Chapter 8:
Productivity in Isolation

In the secluded realm of remote work, personal productivity emerges not just as a necessity but a craft honed in solitude. It's a domain where challenges to stay motivated are as persistent as the glow of a computer screen, yet within this isolation lies an opportunity to redefine what it means to be productive. The absence of physical cues, bustling offices, and the immediate supervision of leaders pushes professionals to forge their own path, relying on self-discipline and innovative techniques to remain efficient. The key is mastering the delicate balance between flexibility and structure; carving out spaces, both mental and physical, where focus can flourish without the usual distractions. As we embrace this solitude, it's crucial to craft a daily rhythm that fuels passion rather than depletes energy, turning isolation into a wellspring of inspiration and creativity. With thoughtful reflection and intentional practice, remote work becomes a gateway to not only meet but exceed one's potential, transforming isolated hours into a mirror reflecting personal and professional growth.

Overcoming Productivity Challenges

As we navigate the complexities of remote work, one of the persistent challenges professionals face is maintaining productivity in isolation. It's not just about staying busy; it's about sustaining momentum and achieving meaningful progress without the immediate buzz of a conventional office environment. Overcoming these challenges requires a multi-faceted approach, combining technology, discipline, and emotional intelligence. Embracing flexible schedules and personalised workspaces can transform how we function daily. Additionally, rethinking the way we set goals and measure success helps adapt productivity principles to fit this new paradigm. Regularly reflecting on achievements and re-evaluating priorities keeps distractions at bay, ensuring focus remains sharp. In this evolving landscape, building robust support networks can inspire consistency and spur motivation. When the line between personal and professional blur, intentionally crafting a structured routine becomes a beacon for productivity amidst isolation. Indeed, the shift to remote work makes us rethink traditional productivity, illuminating paths to harness the full potential of working from home.

Techniques to Boost Efficiency In the journey of navigating productivity in isolation, particularly from the perspective of overcoming productivity challenges, one finds that the critical key is not just working harder, but working smarter. Embracing an adaptive mindset and leveraging effective techniques can truly transform the solitary work experience into a realm of higher productivity and satisfaction. Here, we delve into a range of strategies aimed at enhancing efficiency in remote work settings.

Setting clear boundaries between work and personal life is fundamental. Without the physical separation of an office, the risk of work seeping into personal time increases. Establish a dedicated workspace to signal your brain that it's time to work. This doesn't always have to be a separate room; even a small corner can work wonders. Simultaneously, communicate your work schedule to household members to minimise interruptions. This clarity reduces mental load and boosts focus during work hours.

Another vital technique is harnessing the power of synchronised tools. In isolation, the tools you choose can either be a bridge or a barrier to effective productivity. Choose applications that allow seamless integration of tasks, calendars, and communication. Automating routine tasks—like email sorting, file organisation, and deadline reminders—frees up mental bandwidth for more creative and complex tasks, fostering a proactive rather than reactive approach to work.

Time management, often the Achilles' heel in productivity, becomes crucial. Implement time-blocking techniques where you allocate specific periods for different tasks. This not only aids focus by reducing task-switching but also creates a structured routine that mimics an office environment. Coupled with this, taking regular breaks helps reset your mental state, increasing overall output. The Pomodoro Technique, which involves short, timed intervals of focused work followed by a break, can be highly effective.

Leveraging technology also means learning to collaborate asynchronously. Embrace the concept of asynchronous communication—where not everyone responds in real time, allowing team members in different time zones or with varied schedules to contribute efficiently. Make use of shared documents, recorded video updates, and detailed written communications. These strategies promote uninterrupted work periods and help maintain workflow continuity.

In any remote working scenario, cultivating self-awareness is paramount. Understanding your peak productivity periods—the times when you feel most alert and focused—can significantly enhance your work efficiency. Schedule demanding tasks for these peak times and reserve less intensive tasks for when energy levels dip. Equally, learn to recognise signs of burnout. Regular self-check-ins to gauge stress and motivation levels can prevent work-related fatigue and maintain a steady pace of productivity.

Equally important is the practice of intention-setting at the start of each day or week. Clearly defined goals provide direction and purpose, helping to prioritise tasks effectively. When faced with a heavy workload, it helps to distinguish urgent tasks from important ones using frameworks like Eisenhower's Box. With a purposeful focus, professionals can navigate through tasks systematically, ensuring time is spent on what truly matters.

When it comes to boosting efficiency, consider the role of nutrition and physical health. Although seemingly unrelated, a balanced diet and regular exercise significantly influence cognitive function and energy levels. Being mindful of what you eat, staying hydrated, and incorporating movement into your daily routine—even if it's just a short walk around the house—can considerably improve concentration and stamina throughout work tasks.

While working in isolation, don't overlook the benefits of continual learning. Investing time in skill development not only offers personal growth but also keeps the mind sharp and engaged. Online courses, webinars, and virtual workshops tailored to your field can keep you at the forefront of industry trends and best practices, improving efficiency as a result.

Creating a feedback loop is another key technique. Regularly review your work processes and outcomes to identify areas for improvement. Gathering feedback from colleagues or supervisors helps refine your approaches and ensures you're set on an efficient path. Even informal check-ins can provide valuable insights that drive productivity improvements.

Finally, be sure to remember the power of motivation and self-motivation in a remote work environment. Maintaining a positive mindset and setting personal rewards for task completion can enhance engagement. Visualising success and celebrating small wins can build momentum, propelling you forward with increased enthusiasm and efficiency.

By implementing and adapting these techniques, professionals can not only overcome the productivity challenges of remote work but thrive within them. As the remote work landscape continues to evolve, these strategies will be the bedrock upon which successful, sustainable productivity in isolation can be achieved.

Chapter 9:
Remote Work and
Career Development

In the modern landscape of remote work, career development unfolds with unprecedented flexibility, yet demands a proactive approach. Professionals are discovering that traditional paths to advancement are being reshaped by digital opportunities that reward adaptability, continuous learning, and innovative thinking. Remote work eliminates geographical constraints and reveals career ladders adorned with diverse skill-enhancement avenues and virtual networking possibilities. It's a dynamic realm where individuals can carve out unique career trajectories by leveraging online courses, webinars, and industry-specific platforms to amplify their expertise. Embracing remote work as a catalyst for growth, professionals are inspired to seize control of their development, foster resilience, and channel their newfound autonomy into strategic career advancements. The challenge, now, is not just to adapt, but to thrive in a world where the journey of career progression is no longer linear, but enriched with multifaceted opportunities."

Navigating Career Growth

In the ever-evolving landscape of remote work, charting a course for career growth can feel like navigating uncharted waters. It requires professionals to adopt a mindset that embraces adaptability and relentless learning. Remote settings offer unique opportunities for growth, but they also demand a shift in how we perceive career progress. Rather than relying solely on traditional pathways, individuals must take initiative, proactively seeking out new skills and experiences that align with their ambitions. Building a robust network virtually and showcasing value through digital platforms can open doors that once seemed unreachable. By harnessing the distinct advantages of remote work—such as access to global resources and the ability to craft personalised work environments—professionals can achieve remarkable career trajectories, defying conventional constraints and inspiring newfound resilience.

Skill Enhancement and Learning Opportunities have always been foundational to career development, but the remote work revolution has redefined both the avenues and the pace at which professionals can grow. In the virtual workplace, career progression is no longer tied to the visibility within physical office spaces. Instead, it's about seizing opportunities that may not have been as accessible before. Structure your learning around developing competencies that not only align with your career goals but also adapt to the evolving dynamics of remote work. How can you equip yourself for unforeseen challenges and seize future job opportunities?

The journey of skill enhancement in a remote setting begins with embracing a mindset of perpetual learning. With technology removing geographical limitations, professionals now have access to a treasure trove of online courses, webinars, and workshops that cater to a diverse array of skills and industries. You might explore platforms like Coursera, Udemy, or even company-provided training modules to delve into new territories or deepen existing expertise. The flexibility of remote work means you can chart out a personal development plan that fits seamlessly into your daily routine, blending work and learning harmoniously.

While self-directed learning is valuable, don't overlook the significance of mentorship in your career development journey. Mentors provide a nuanced perspective on navigating industry waters and offer guidance that's often far removed from textbook knowledge. Remote work has inadvertently expanded the mentor pool. You're no longer restricted to seeking mentors within your physical office or locality. Instead, connect with experienced professionals worldwide who resonate with your career aspirations. Establish virtual coffee chats, schedule regular check-ins, and create lasting professional relationships that enrich both your personal and career growth.

In parallel to finding mentors, consider the power of building informal networks or learning communities. Online groups and forums dedicated to your specific industry offer a platform to share insights, discuss trends, and troubleshoot challenges collectively. Platforms like LinkedIn and professional Slack channels can be instrumental in this quest. Engage in discussions and build your reputation. Moreover, your contribution to these communities can also flag you as a leader in your field, opening the door to collaboration opportunities and career advancement.

Practical application of newly acquired skills is crucial for career growth in a remote environment. The ideal scenario is to integrate your fresh knowledge into your current role. This not only benefits your organisation but also showcases your adaptability and forward-thinking approach. Are there projects where you can contribute your newfound expertise? Put your hand up for cross-departmental tasks or global initiatives that can leverage your skills. Furthermore, remote work encourages innovation – experiment with innovative solutions to problems that emerge in the virtual work sphere, and don't be afraid to take calculated risks.

Feedback remains a pillar of career development. In remote work, proactive communication about your progress, challenges, and feedback is vital. Regular one-on-one meetings with supervisors and peers can act as sounding boards for your career growth trajectory. Use these sessions to discuss training opportunities, evolve role expectations, and receive constructive criticism. The path of remote career development demands that you sometimes step out of your comfort zone to initiate these conversations, but the rewards, in terms of personal and professional growth, are invaluable.

Remote environments also necessitate an even sharper focus on developing 'soft skills', as these can significantly impact career growth. Effective communication, time management, self-discipline, and digital etiquette are crucial in a remote setting. Unlike hard skills, these are not easily measured but are equally important in your remote career toolkit. How you convey your ideas in a video call, manage your projects across time zones, and show empathy in written communication all define your growth and success in a remote landscape.

The nature of collaborative work has changed, too, requiring more agility and adaptability. Virtual collaboration tools have made geographical differences almost negligible. However, the need for technical proficiency in these tools alongside the ability to work inclusively and efficiently in a diverse team is paramount. Employers treasure team members who are adept at leveraging collaborative technologies to drive productivity without losing the essence of teamwork.

The landscape of remote work presents a striking opportunity to redefine traditional career progression hierarchies. It's a time to recognise the potential of lateral movements and skill diversifications as valid steps in career enhancement. Rather than a linear path upward, guidance and learning opportunities should include the exploration of different facets of one's role or industry. This approach not only prevents occupational stagnation but also prepares professionals for a broad range of responsibilities.

Keep a close eye on industry trends and disruptions. The remote work setup can sometimes cause a disconnect from industry happenings if proactive efforts are not made to stay informed. Attending virtual seminars, subscribing to relevant newsletters, and engaging with thought leaders on social media platforms are effective ways to remain current. Knowledge of industry trends will not only enhance your skill set but also position you as a knowledgeable resource within your organisation.

Skill enhancement in remote work bridges professional development with the inherent human desire for self-improvement. It's about carving out paths in uncharted territories, powered by technology and driven by personal ambition. Navigating career growth in this era calls for an innovative approach to learning – one that leverages the wide-open access to global resources while maintaining the dedication to personal and professional progress.

Finally, don't underestimate the value of reflection in your learning journey. Periodically evaluate your growth – are your efforts aligned with your career goals, or is there a need to pivot? This reflection will keep you anchored and focused. Remember, the remote work environment provides an enriching platform for those brave enough to embrace the endless learning opportunities it affords.

Chapter 10:
Managing Remote Conflicts

As remote work reshapes the professional landscape, navigating conflicts in this virtual arena demands a fresh approach. With screens separating colleagues, the subtle cues of body language and the immediacy of physical presence are lost, complicating resolution efforts. Yet, the very nature of remote environments offers a unique opportunity to tackle disputes with empathy and innovation. Embracing clear communication channels and harnessing technology can transform potential conflicts into opportunities for growth and understanding. Leaders and employees must develop an acute awareness of cultural differences and diverse working styles, fostering an environment where diverse voices are not just heard but valued. By promoting open dialogue and using creative conflict resolution techniques, teams can convert friction into fuel for collaboration, fostering a resilient and harmonious remote workspace. The challenge lies in not just addressing conflicts as they arise but in cultivating relationships that are strong enough to withstand misunderstandings and foster collective progress.

Addressing Disputes Virtually

In the expansive landscape of remote work, addressing disputes virtually stands as both a challenge and an opportunity for modern professionals. Conflicts now emerge not in boardrooms or office corridors, but in digital spaces that, while convenient, lack the nuances of face-to-face interaction. Resolving these disputes requires more than just adeptness with technology; it demands a renewed focus on empathy, active listening, and nuanced communication strategies. Virtual platforms, from video calls to instant messaging, can exacerbate misunderstandings, yet they also offer unique tools for navigating conflicts effectively. By fostering open dialogue, encouraging transparency, and leveraging technology innovatively, professionals can transform disputes into constructive conversations that foster growth and understanding. This approach not only mitigates the immediate conflict but also builds a foundation of trust and cooperation that elevates entire teams. In embracing these methods, leaders and individuals alike can redefine how we perceive conflict resolution in the digital age, enhancing collaboration and pushing the boundaries of what is possible in our increasingly virtual workplaces.

Conflict Resolution Techniques delve into the delicate art of finding harmony in a world where physical cues and face-to-face interactions are often absent. As we explore managing remote conflicts, addressing disputes virtually becomes a crucial part of ensuring a cohesive and productive work environment. It's about blending empathy with technology to drive solutions that not only solve issues but also strengthen team cohesion.

Virtual settings, while offering unprecedented flexibility, can also be breeding grounds for misunderstandings. Imagine trying to discern the tone of a curt email or a missed tone in a video call—without the usual nonverbal signals, the propensity for conflict increases. Thus, conflict resolution techniques must adapt to this new terrain, transcending traditional methods to be effective in a dispersed work environment. Understanding the subtle nuances of remote interactions is essential.

Central to resolving conflicts remotely is communication clarity. Clear, concise, and direct communication can prevent many conflicts before they escalate. In a virtual space, transparency and precision are key. Encourage team members to be explicit about expectations and timelines, reducing the room for ambiguity. When everyone is on the same page, there's less likelihood of misinterpretation turning into a dispute.

Active listening emerges as another cornerstone technique. In virtual settings, this involves not just hearing the words but paying attention to the underlying sentiment. Facilitating spaces where everyone feels heard and valued promotes a healthy team dynamic. Encouraging team members to echo back what they've understood during discussions can ensure alignment and understanding, thereby nipping potential conflicts in the bud.

Empathy, though challenging to convey across digital mediums, remains indispensable. Leaders must nurture an environment where empathy isn't just encouraged but practised. This means taking the time to understand the diverse circumstances each team member faces—be it a barking dog in the background or a sudden home emergency. Recognising these challenges and showing genuine concern fosters trust and decreases tensions.

Technology, often the source of miscommunication, doubles as a mediator in conflict resolution. With the advent of AI-driven tools, real-time feedback and sentiment analysis can help gauge the mood of conversations and identify potential conflict points early. More sophisticated platforms offer virtual mediation sessions where involved parties can engage with trained professionals to guide the resolution process.

Structured conflict resolution frameworks adapted for the virtual environment can be incredibly effective. Popular models like the "Interest-Based Relational Approach" or the "Thomas-Kilmann model" can be tweaked for remote application. Having a set procedure provides a roadmap for teams to follow during disputes, ensuring a consistent approach that everyone is familiar with and agrees upon.

Feedback loops are vital too. They offer a systematic channel to voice concerns and raise issues before they fester into conflicts. Regular feedback sessions, where team members can express concerns in a moderated setting, contribute to a more harmonious workplace. These feedback loops should be seen as opportunities for growth, fostering a culture of continuous improvement rather than criticism.

One cannot underscore enough the importance of documenting agreements and resolutions. In virtual settings, where fleeting verbal assurances can be forgotten, formalising resolutions in writing can prevent future disagreements. This documentation serves as a reference point, reminding all parties of the commitments made and the path forward.

Training plays a crucial role in equipping team members with the skills necessary for effective virtual conflict resolution. Regular workshops focusing on digital communication etiquette, emotional intelligence in virtual settings, and advanced negotiation strategies can empower teams to handle disputes independently and maturely.

Building a strong, conflict-resilient culture starts with leadership. Leaders should model conflict resolution behaviours they wish to see. When leaders openly navigate their conflicts gracefully and constructively, it sets a precedent for the rest of the team. Encourage leaders to be approachable, where discussions on disagreements are viewed as opportunities rather than threats.

Lastly, scheduling regular team-building activities can alleviate tensions. These aren't just social occasions; they're strategic opportunities to strengthen interpersonal relationships and build empathy among team members. Virtual retreats, collaborative projects, and even simple coffee breaks over video calls can cultivate a sense of belonging and solidarity, creating a safety net of understanding that cushions the impact of conflicts when they do arise.

In conclusion, mastering conflict resolution techniques in a virtual environment demands a blend of old wisdom and new strategies. It's about acknowledging the unique challenges remote work presents while leveraging the tools and resources available to address these. By fostering clear communication, active listening, empathy, and structured processes, professionals can navigate conflicts constructively, maintaining harmony and productivity in the virtual workplace.

Chapter 11:
Legal and Ethical Considerations

As the shift towards remote work accelerates, navigating the legal and ethical terrain becomes imperative for organisations and professionals alike. Businesses must grapple with a labyrinth of compliance requirements and data security regulations to safeguard sensitive information, all while balancing ethical considerations that ensure fairness and transparency. This transition calls for robust policies that protect both the organisation and its remote workforce, fostering trust and a sense of duty. Leaders and managers face the challenge of developing a framework that aligns with legal mandates, respects employee privacy, and embraces the evolving digital landscape. As we chart this course, our commitment to ethical practices becomes the compass guiding us towards a future where innovation thrives within the bounds of law and integrity.

Compliance and Data Security

In the ever-evolving landscape of remote work, compliance and data security stand as towering pillars in safeguarding both organisational integrity and individual privacy. As professionals traverse this digital terrain, they're challenged to integrate robust security measures seamlessly, ensuring that data remains impenetrable while fostering an environment of trust. Organisations and individuals alike must navigate an intricate web of regulations, applying a proactive approach to mitigate risks and embrace best practices that transcend geographical boundaries. It's not just about meeting regulatory needs; it's about cultivating a culture where security becomes second nature, inspiring confidence and enabling innovation without borders. As we advance into this modern work era, let's champion a future where compliance is not merely an obligation but a blueprint for sustainable growth and ethical accountability.

Policies to Protect Organisations As organisations increasingly settle into the rhythm of remote work, the necessity for robust policies to shield against potential legal and ethical pitfalls becomes glaringly evident. Protecting the organisation extends beyond just ensuring data security; it's about crafting a framework that facilitates compliance while respecting the rights of employees and stakeholders alike. Effective policies are, in essence, the backbone that supports an enterprise's foray into remote operations, providing clarity and instilling confidence.

First and foremost, an organisational policy aimed at protection ought to encompass comprehensive data protection strategies. Since remote work invariably involves extensive digital communication, businesses must establish protocols that govern data handling and storage. This isn't merely about meeting legal requirements but also about cultivating a culture of trust where employees and clients feel their information is safeguarded. Encrypting data, updating software regularly, and investing in secure cloud storage solutions are all measures that fortify the organisation's digital defences.

Furthermore, compliance with regional and international regulations is not optional; it's a necessity. Different geographies come with distinct legal landscapes. For instance, the General Data Protection Regulation (GDPR) in Europe sets a high bar for data privacy that companies must adhere to, irrespective of their location, if they deal with EU residents' data. Therefore, policies must be agile, accommodating legislative updates and adapting proactively to regulatory changes. Employers can benefit from consulting legal experts to ensure that policies are both comprehensive and compliant without being overly intrusive.

Equally important is the need to elucidate the expectations and responsibilities of remote employees. Employees should be made aware of what constitutes acceptable use of company resources, such as laptops and software. Clear guidelines minimize the risk of data breaches caused by negligence. Training sessions and regular refreshers can empower workers by highlighting the importance of maintaining cyber hygiene, thus fostering an informed workforce.

But what of fostering an ethical remote working environment? Beyond legal boxes to tick, it's crucial that companies uphold ethical standards that protect both the organisation and the individual. This involves creating policies that deter unethical practices such as remote micromanagement, which could encroach on personal privacy and lead to a toxic work environment. Policies should uphold transparency and contain provisions for anonymous whistleblowing mechanisms, ensuring that employees can report unethical behaviour without fear of retribution.

Moreover, organisations should also consider the ergonomic well-being of their remote staff. Policies might include stipulations for home office setups or reimbursement programs for essential equipment. This isn't solely about productivity; it addresses legal and ethical responsibilities concerning employee health and safety. After all, a healthy, supported worker is a productive and engaged one.

In addition, there's a pressing need to define contingency protocols within these policies. Cyber-attacks are a real threat that can derail operations. Policies should outline the steps to take in the event of a breach, designating responsibilities and communication strategies during such crises. Regularly conducted mock drills can prepare an organisation to respond effectively, minimising potential damage.

Developing a policy is one thing; implementing it effectively is another. To truly protect their interests, organisations need to ensure these policies are visible and accessible. An easily navigable digital handbook that's regularly updated and disseminated can serve as a great repository of these policies. The goal should be to make compliance a facet of everyday operations rather than an annual checkbox.

Collaboration is pivotal in designing these policies. Involve stakeholders across different functions, from IT to HR, in the drafting process. Each department may have unique insights into potential risks and the necessary countermeasures. This collective approach not only enhances the policy's comprehensiveness but also secures buy-in from the various sectors of the organisation. Moreover, it fosters a sense of ownership and accountability, as each participant offers their expertise to shape the organisational framework.

Lastly, measuring the effectiveness of these policies is imperative. Establish KPIs related to data protection and compliance, such as the number of data breach incidents or the completion rate of compliance training modules. Regular audits can help identify gaps and areas for improvement, facilitating continuous enhancement of the organisation's protective protocols.

In conclusion, policies to protect organisations are more than a legal requirement; they are the safety net that allows innovation and remote working practices to flourish unhindered by unnecessary risks. By placing equal emphasis on compliance and ethics, organisations can create an environment that supports growth and agility while maintaining the trust of the workforce and clientele alike. The future of work leans heavily on how well these policies are crafted and integrated, illustrating the pivotal role they play in today's ever-evolving professional landscape.

Chapter 12:
Remote Work and Diversity

In the ever-evolving landscape of remote work, diversity not only enriches teams but becomes a cornerstone for innovative solutions and cohesive unity across borders. The virtual work environment levies a unique opportunity to cultivate a workforce that truly represents a mosaic of backgrounds, experiences, and perspectives. Championing inclusivity in remote settings is not merely a checkbox for HR departments but a strategic advantage that intertwines empathy with creativity. By tapping into the talents of a diverse pool unbound by geographic constraints, organisations can thrive on fresh ideas and nuanced understanding, ultimately leading to exponential growth. Encouraging managers to actively dismantle barriers and foster inclusive dialogues ensures voices often sidelined are heard and valued. The challenge lies in translating traditional diversity strategies into a seamless digital experience that not only celebrates individual uniqueness but forges an interconnected web of shared goals. In doing so, we aren't just reshaping organisational ethos, we're redefining what it means to be truly global. Let's embrace diversity not as a chapter in our remote work handbook but as the lens through which we see and craft the future of work itself.

Enhancing Inclusivity

In the remote work era, enhancing inclusivity isn't just an ideal; it's a mandate. As physical barriers dissolve, new opportunities arise to embrace a truly global workforce. Yet, it's vital to acknowledge and address the nuanced challenges of equitably engaging a diverse team spread across different cultures and time zones. The common platform of remote work provides a unique chance to level the playing field, giving voice to underrepresented groups by valuing diverse perspectives. By fostering an inclusive virtual environment, organisations can tap into a broader range of ideas, experiences, and innovations, driving success in unprecedented ways. Investing in inclusivity is not merely a box to check but a strategic advantage that can propel teams towards a richer, more collaborative future. Through thoughtful policies and sincere commitment, remote work can transform the way we define inclusivity in the digital age.

Creating a Diverse Remote Environment In the rapidly evolving landscape of remote work, fostering a diverse environment isn't merely a desirable goal; it's an absolute necessity. As organisations strive to enhance inclusivity, it's critical to understand that diversity transcends mere demographics—it involves embracing a multitude of perspectives, experiences, and ideas. By shaping a virtual workspace that respects and celebrates these differences, companies can tap into a rich tapestry of creativity and innovation that propels them forward in a competitive world.

Central to creating a diverse remote environment is the robust implementation of inclusive practices at all levels of an organisation. It's about more than fulfilling quotas or appearing equitable; it's about unlocking the full potential of a diversified workforce. In remote settings, where geographical barriers are minimised, the opportunities for diversity are boundless. However, these opportunities must be accompanied by intentional strategies to ensure that all voices—no matter how distant—are heard and valued.

One strategy involves implementing equitable recruitment processes. Companies should expand their search for talent beyond traditional borders, leveraging technology to reach underrepresented groups. They must consider candidates from various backgrounds, recognising that diverse life experiences contribute uniquely to the workplace. Ensuring that hiring panels themselves are diverse can mitigate unconscious biases, promoting a fair assessment of potential hires.

Once talent is onboarded, nurturing an environment where everyone feels a sense of belonging is crucial. Regularly facilitating open dialogues about diversity helps establish a culture where employees feel safe expressing their concerns and ideas. Remote platforms can be harnessed to create forums for these discussions, whether through structured meetings or informal coffee chats. The key is fostering an atmosphere where each team member knows their input is valued and impactful.

Building a diverse remote environment also involves providing equal access to resources, opportunities, and support. Organisations should ensure that all employees have the necessary tools to succeed, regardless of their location or background. This can involve providing technology to those in areas with less access or offering training to enhance digital literacy. It's about ensuring that everyone starts from a level playing field and has the chance to grow and contribute meaningfully.

Leadership must play an active role in driving and sustaining diversity initiatives. By embodying inclusive values, leaders set the tone for the entire organisation. This involves a commitment not just to policy changes but to a sustained cultural shift. Leaders should be champions of diversity, actively seeking out and valuing different perspectives in decision-making processes. Their ability to show vulnerability and a willingness to learn and adapt is essential in creating a truly inclusive environment.

Mentorship and sponsorship are also vital in promoting diversity. Remote settings can inadvertently obscure the visibility of minority voices, so it's crucial for organisations to implement systems where employees can find mentors who guide and advocate for their career advancement. These relationships can inspire confidence and provide critical insights, helping individuals navigate the complexities of remote work with a strong support network.

Moreover, organisational structures need to be reflective of diversity and equity. This might include task forces or committees dedicated to monitoring diversity goals, which are equipped with the power to influence change. By regularly reviewing diversity metrics, companies can ensure they stay aligned with their goals, adjusting strategies as needed to address any gaps or challenges.

Simultaneously, encouraging cross-cultural collaboration can be extremely beneficial. Leveraging the geographical spread of remote teams, companies can create global project teams that unite diverse perspectives. This not only enriches the work produced but also enhances cultural appreciation and understanding amongst team members.

Reflecting diversity in company communications is another lever. This means showcasing diverse voices and narratives through newsletters, webinars, and other platforms. By regularly highlighting diverse contributions, organisations reinforce the message that diversity is not only welcome but essential to their identity.

Finally, measuring success in creating a diverse remote environment entails more than just quantifiable metrics. Companies must also gauge the qualitative impact of their diversity initiatives. This involves seeking feedback from employees and being ready to listen and adapt based on their lived experiences. A remote environment that feels inclusive and empowering is a sure sign of success, one where employees, irrespective of identity, feel they belong and can thrive.

In conclusion, creating a diverse remote environment requires a multifaceted approach that combines recruitment, culture, support, and leadership. It's not a one-off initiative, but an ongoing journey that demands commitment and creativity. As organisations embrace these principles, they do more than enhance inclusivity—they unlock the full potential of their workforce, crafting a future that's vibrant, dynamic, and equitable for all.

Chapter 13:
Innovation and Creativity in Isolation

Amidst the solitude of remote work, a paradox emerges—the very isolation that separates us can also serve as a crucible for innovation and creativity. Without the confines and conventions of traditional office spaces, professionals find themselves in an uncharted terrain where old routines give way to fresh perspectives. It's here in the quiet of one's own space, liberated from the ever-watchful office setting, that creative resolutions and groundbreaking ideas often germinate. This isolation provides the fertile ground for introspection and uninhibited thought, allowing ideas to percolate and mature unencumbered by typical workplace distractions. When nurtured, these ideas coalesce into innovative solutions that propel businesses forward in novel directions. Employers must strive to foster environments that encourage playful exploration and reward out-of-the-box thinking, even at a distance. By embedding trust and autonomy into remote cultures, leaders can unlock the immense creative potential inherent in their teams, catalysing a new era of invention defined not by proximity, but by purpose and imagination.

Encouraging Creative Thinking

In the seemingly solitary confines of remote work, fostering creativity can feel like lighting a match in the wind, yet it's precisely this solitude that acts as a catalyst for innovation. When professionals are distanced from traditional workplace norms and interruptions, they find themselves with the space to think divergently. Encouraging this kind of thinking lies in creating a digital environment that embraces flexibility and curiosity. Companies can nurture a culture where risk-taking is celebrated, and walls don't define boundaries. Virtual brainstorming sessions, unpressured by office time clocks, can ignite inventive solutions, while collaborative tools ensure these ideas aren't lost in the ether. Establishing a sense of psychological safety in online platforms allows team members to share audacious ideas without fear of judgment. By reimagining isolation not as a barrier but as a canvas, organisations and individuals alike can tap into a wellspring of creative potential, turning remote work into a hotbed for fresh perspectives and breakthrough ideas.

Mechanisms to Spur Innovation In the evolving landscape of remote work, fostering innovation is not just a corporate buzzword — it's a necessity for survival and growth. As professionals navigate the unique challenges of isolation, the importance of cultivating an environment where creative thinking can flourish becomes paramount. This is not merely about having innovative ideas, but about creating processes and conditions where these ideas can be born and nurtured.

One of the fundamental strategies to spur innovation is redefining how teams approach problem-solving. Traditional brainstorming sessions, once confined to physical boardrooms, need to adapt to the virtual world. This transition, while challenging, presents opportunities that are often underestimated. Virtual ideation methods, such as digital whiteboards and collaborative platforms, offer diverse avenues for spontaneous inspiration, giving all team members an equal voice and breaking down hierarchical barriers to creativity.

Moreover, encouraging creativity in isolation requires a culture that actively promotes experimentation and accepts failure as part of the learning process. When teams work remotely, the fear of failure can be exaggerated due to the lack of physical presence and immediate support. Therefore, leaders must communicate and demonstrate that innovation often emerges from a series of iterative failures before achieving success. This approach not only lowers the barriers to creative risk-taking but also instills resilience within teams.

In addition, harnessing the power of diversity can act as a catalyst for innovation. Remote work offers unique opportunities to bring together people from varying backgrounds, cultures, and perspectives, creating a melting pot of ideas that can lead to groundbreaking solutions. Encouraging cross-departmental collaborations and international project teams can infuse fresh perspectives into problem-solving and spark creativity that transcends organisational boundaries.

Another potent mechanism involves leveraging technology to automate mundane tasks, freeing up the mental bandwidth of team members to focus on more complex, innovative tasks. When workers are bogged down with repetitive activities, their capacity for creative thinking shrinks. Implementing AI-driven tools and software to handle routine processes can significantly enhance the capacity for innovation by allowing employees to engage more deeply with their work and think outside the box.

A critical but often overlooked factor is the role of the physical environment, even in a remote setup. While individuals might not be in a traditional office, their home workspace can greatly influence their capacity for creativity. Providing guidance and support for remote workers to design their optimal workspaces — spaces that inspire rather than stifle — can have a profound impact on their ability to innovate. This might involve offering subsidies for ergonomic furniture or providing access to online workshops that teach creativity-boosting techniques, such as mindfulness or decluttering strategies.

Organisations must also invest in continuous learning opportunities. In the fast-paced world of remote work, skills can quickly become obsolete, and innovation stalls without fresh knowledge. Offering a plethora of online courses, webinars, and workshops in emerging fields not only keeps skills fresh but also immerses employees in diverse knowledge pools that can trigger innovative thoughts and solutions.

Creating 'creative clusters' or innovation hubs within virtual teams can also drive transformative ideas. These clusters could function as small, task-oriented groups tasked with exploring new ideas or solving specific challenges. By rotating these groups and involving different team members, organisations can glean varied insights and foster a consistent pipeline of innovative thinking.

Another strategy that's gaining traction is the 'hackathon' model, adapted for the remote work environment. Virtual hackathons are time-boxed events where individuals or teams intensively collaborate to solve a problem or develop a project. These events can ignite innovation by temporarily setting aside daily duties, freeing time to tackle new challenges, and fostering a spirit of competition and reward.

Effective communication remains a linchpin in driving innovation. Creating clear, open channels for dialogue ensures that ideas can flow freely without hindrance. Establishing regular check-ins and brainstorming sessions where all voices are heard can prevent innovative suggestions from being lost in the vastness of virtual communication. Real-time feedback loops help to refine ideas quickly, maintaining momentum and focus.

Finally, recognising and celebrating innovation can provide significant motivation and inspiration for individuals and teams. When innovative achievements are publicly acknowledged and rewarded, it sends a powerful message to the entire organisation: creativity is valued and rewarded. This recognition doesn't just motivate the award-winning individuals but encourages others to strive for creative excellence.

As professionals continue to adapt to the remote work era, these mechanisms provide a roadmap for sparking innovation in isolation. By redefining processes and embracing the opportunities afforded by technology and diversity, organisations can not only survive but thrive in this new landscape. Creativity needn't be a casualty of remote work; with the right strategies, it can instead be its greatest gift.

Chapter 14:
Remote Work's Impact on Real Estate

The advent of remote work has transformed not only our daily routines but also the very spaces we inhabit. As many professionals embrace the flexibility of working from anywhere, the demand for traditional office spaces is being re-evaluated. This shift has led to a reimagining of urban landscapes as companies reconsider the necessity of towering office blocks, opting instead for more adaptable and multifunctional spaces. Meanwhile, residential real estate is witnessing a pivot towards homes equipped with dedicated work environments, making home offices a sought-after feature. The ripple effect extends beyond urban borders, as rural and suburban areas gain attractiveness due to the desire for larger living spaces and a better quality of life, free from the constraints of daily commutes. For real estate stakeholders, understanding these trends is crucial in anticipating the future needs of a workforce that balances flexibility with functionality, reshaping the very foundation of how and where we choose to live and work.

Shifts in Housing and Office Spaces

The seismic shift to remote work is reshaping our relationship with both housing and office spaces, blurring lines that once seemed etched in stone. As city centres empty and people flee to suburbs or rural idylls in search of a new work-life nirvana, the ripple effects are profound. Demand for spacious homes with designated work areas is spiking, and this creates a buyer's market far from urban sprawl. Meanwhile, corporate giants grapple with the reimagining of office environments — transitioning lavish headquarters into flexible, collaborative hubs or even reducing footprints entirely. What was once the heart of business is transforming, urging us towards a hybrid model where office visits become purposeful rather than routine. In this dance of adaptation, traditional real estate models are compelled to evolve; remote work doesn't just challenge them, it redefines their core. Ultimately, how we live and work is shifting into something more fluid, with opportunity blooming wherever adaptability thrives.

Future Real Estate Trends will emerge as remote work continues to reshape the way we live and utilise spaces. The pandemic catalysed a massive shift towards working from home, leading to an evolution in both housing and office dynamics. As this change becomes more entrenched, professionals at every level—from HR leaders to corporate managers—must grapple with the real estate implications of this new way of working. What does the future hold for spaces where people live and work? How can we best adapt to these shifts in housing and office spaces that are rapidly becoming the new norm?

One of the most profound trends is the de-urbanisation of residential areas, spearheaded by the increased flexibility remote work affords. With no daily commute binding employees to urban centres, there's a growing appeal in suburban and even rural settings. People are moving away from crowded city apartments to houses that offer more space, both indoors and out, enabling a lifestyle that aligns with remote work requirements. This trend is not just a temporary phenomenon; it represents a significant cultural shift in living preferences. Larger homes to accommodate home offices and a stronger desire for outdoor spaces have become critical criteria in the housing market.

Work-from-home policies have led firms to reassess the need for large office spaces in city centres, resulting in a decline in demand for traditional office real estate. This has sparked innovative uses of space, pivoting towards designs that accommodate flexible office hours and hybrid work models. Companies are reimagining office environments not as places of daily drudgery but as hubs for collaboration and creativity when physical presence is needed. The office of the future is likely to blend co-working spaces with traditional amenities while fostering a sense of community among remote workers who seek face-to-face interaction occasionally.

This shift has wrought significant implications for commercial landlords who now face the challenge of repurposing office buildings. Adaptive reuse of structures is becoming a catchphrase in the real estate world, with buildings being transformed into mixed-use spaces, combining residential, commercial, and even recreational facilities. This trend not only meets the changing demands but also revitalises urban areas that might otherwise suffer from soaring vacancy rates. Additionally, businesses are considering spaces that offer fewer long-term commitments to allow flexibility in uncertain economic landscapes.

Technology plays a crucial role in these transformations, enabling seamless integration between one's home and workspaces in remote settings. Smart home technologies are becoming more prevalent, allowing employees to create efficient and productive environments. From high-speed internet to smart office setups, the future of real estate will increasingly integrate technology to meet individuals' remote working needs. This is driving a demand for infrastructure improvements in broadband and wireless technologies, particularly in suburban and rural regions.

Moreover, the concept of community is being reshaped by the emergence of neighbourhood work hubs. These are shared spaces located within residential communities that allow remote workers to interact and collaborate locally, offsetting the isolation that can accompany prolonged remote work. Such hubs provide the social interaction necessary for productivity and well-being, while also fostering local engagement and potentially reducing the carbon footprint associated with commuting.

In terms of commercial real estate investment, there is a noticeable shift towards residential and logistics-focused properties. Warehouses and distribution centres are thriving in a world where online shopping has exploded, indicating that such investments might offer higher returns than traditional office spaces. For investors, this opens up a plethora of new opportunities in a sector undergoing rapid transformation.

Furthermore, cities themselves are evolving. With fewer people tethered to offices, urban planners and civic officials are rethinking allocations of space within cities. Public spaces, parks, and transport networks are being redesigned to prioritise pedestrians and cyclists over cars, creating greener, more liveable environments. Flexible zoning laws and incentives for sustainable building practices are likely to gain traction as communities think long-term about the fusion of work and home in peaceful and productive coexistence.

The ripple effects of these trends are vast, touching on everything from environmental sustainability to economic revitalisation. Where once office districts would stand empty post-working hours, mixed-use developments promise round-the-clock vibrancy. Yet, this transition requires foresight and planning. Corporate managers, HR leaders, and even policymakers must be proactive in accommodation, evolving employee needs within strategic perspectives on urban development.

This confluence of remote work and real estate heralds a future where flexibility, technology, and sustainability align to create spaces that meet modern workforce expectations. The journey into these future trends is not merely about keeping pace but paving new roads of opportunity that reflect transformative societal shifts. As the boundaries between work and living change, the landscape of real estate is set to evolve into something far more dynamic and purposeful than ever imagined.

Chapter 15:
Social Dynamics and Remote Work

The shift to remote work has revolutionised social dynamics, reshaping how professionals connect and engage within a virtual landscape. No longer confined by geographical boundaries, workers can tap into a global pool for community interaction, sparking conversations and collaborations that transcend time zones and cultures. While this presents unique opportunities for virtual networking and participation in digital events, it also necessitates a reevaluation of how we nurture a sense of belonging and camaraderie without physical presence. Leaders need to be creative, fostering engagement through thoughtful digital interactions and curated virtual spaces that invite authentic connection. In this brave new world, the blend of professional and personal lives calls for a focus not just on efficiency but also on empathy, encouraging teams to embrace technology as a bridge rather than a barrier. The forward-thinking organisation sees the evolution of social dynamics as a chance to enrich the workplace experience, weaving connectivity into the fabric of remote work to enhance productivity and well-being.

Connectivity and Community Engagement

The shift to remote work has redefined how professionals connect and engage with their communities, both within and beyond their organisations. Connectivity now hinges on virtual windows where meetings replace watercooler chats, and digital platforms transform social interactions into planned engagements. Building connections in this landscape requires intentional action, leveraging technology as a bridge rather than a barrier. By fostering a sense of belonging through shared digital experiences, remote workers can cultivate camaraderie and community spirit even at a distance. Furthermore, creating virtual networks opens doors to diverse perspectives and innovative collaborations, enhancing both professional growth and personal fulfilment. While the lack of physical presence poses challenges, it also unlocks opportunities to forge meaningful connections across geographical boundaries, redefining community engagement in bold and inspiring ways. Embracing these dynamics enables remote professionals to remain integrated while contributing to a vibrant, interconnected world.

Virtual Networking and Events are more than just a convenience in today's remote work environment; they are becoming a cornerstone of effective professional connectivity and community engagement. As remote work reshapes our social interactions, how we network and participate in events must also evolve. Gone are the days when professional networking only occurred in physical spaces such as conferences or seminars. Today, virtual networking affords professionals the ability to connect regardless of geographical barriers, expanding not just our personal, but our shared professional landscapes.

Effective virtual networking isn't just about exchanging contact information; it's about forming genuine connections that can lead to collaborative success. Platforms like LinkedIn, virtual conference software, and community forums have risen as essential tools. These platforms enable individuals to continue discussions and build relationships that were once only possible face-to-face. The virtual realm provides additional methods to maintain and nurture these connections with a flexibility that traditional networking cannot match.

Understandably, the virtual setting can be intimidating for some. The absence of physical cues and the distinct dynamic of online interactions require a recalibration of networking strategies. Participants must now be more intentional in their engagements, utilising clear communication and active listening to forge connections. Mastering the art of virtual body language, such as thoughtful facial expressions and posture within video calls, has become just as crucial as a firm handshake once was.

Moreover, virtual events now encompass a broad spectrum of possibilities, from webinars and online workshops to full-scale conferences and trade shows. These events can range from intimate gatherings of like-minded individuals to expansive events hosting thousands. The capacity to connect globally, sharing diverse perspectives and knowledge, is accelerating innovation and understanding across industries. Yet, the impersonal nature of screens demands unique approaches to engagement.

Hosting a successful virtual event requires meticulous planning and creativity. Organisers must think beyond merely translating in-person frameworks to the digital space. Interactive elements, such as breakout sessions, live polls, and networking lounges, are vital in maintaining participant engagement and fostering a collaborative environment where ideas can flourish. Infusing elements of fun and novelty into these events can transform them from sterile digital meetings to vibrant networking opportunities.

Challenges will arise, of course. Technical difficulties can disrupt even the best-crafted events. Thus, robust tech-support and contingency plans are indispensable. Furthermore, time zone differences may hinder global participation, demanding an inclusive approach to scheduling and possibly offering on-demand content to accommodate all participants. As a result, these challenges urge organisers to aim for flexibility, ensuring that virtual events are accessible, inclusive, and engaging for everyone involved.

Virtual networking and events also pose opportunities for creative leadership within organisations. Leaders can leverage these platforms to break down internal barriers, encourage open dialogue, and strengthen workplace relationships. By promoting knowledge sharing through virtual platforms, companies can create cohesive teams that thrive on collective growth and innovation.

Social capital—the intangible yet quantifiable value of social networks—flourishes in environments that encourage connections and collaborations. Organisations can bolster this capital by fostering environments conducive to both formal and informal virtual networking. Engendering a culture where employees feel empowered to initiate and join professional networks can drive significant benefits, including heightened job satisfaction and improved performance.

In building a remote-friendly culture, companies should encourage participation in virtual industry events, contributing not only to individual professional growth but also to fostering a collaborative industry ecosystem. This not only elevates the individual but also robustly positions the organisation within its sector. The use of virtual reality and augmented reality is emerging as an innovative frontier, offering immersive networking experiences that can simulate real-world interactions more closely.

Ultimately, the way we engage in virtual networking and events is a reflection of the broader transformation in the social dynamics of remote work. It's about strategically blending technological innovations with the timeless human need for connection and community. As we adapt to these changes, a world of opportunities to unite, learn, and grow opens up—far beyond the constraints of physical spaces.

The rise of virtual networking and events marks a significant cultural shift in professional engagement. It's a clarion call for professionals, HR leaders, and corporate managers alike to embrace these changes and champion initiatives that foster meaningful interactions. By doing so, they ensure that remote work doesn't become an isolating experience but a catalyst for connectivity and collaboration, laying a foundation for a continuous and enriching professional journey.

Chapter 16:
Health and Wellbeing
in a Remote Era

As the remote work landscape continues to expand, maintaining health and wellbeing has become an essential focus for professionals and organisations alike. In this new era, blending work and life under one roof necessitates creative solutions for both mental resilience and physical health. With gym spaces often replaced by living rooms and human interactions frequently mediated through screens, the challenge is real. Organisations have a unique opportunity to innovate holistic wellbeing programs that cater to individual needs, integrating flexible schedules, virtual workout sessions, and mental health support. As boundaries blur, the practice of proactive self-care becomes pivotal, enabling employees to cultivate balanced, fulfilling lives despite the absence of traditional office structures. By prioritising health and wellbeing, remote work can transform from a mere survival strategy to a thriving lifestyle choice, leaving both employees and employers empowered and inspired.

Physical Health Considerations

In the burgeoning era of remote work, maintaining physical health isn't just an optional pursuit—it's a cornerstone of sustained productivity and overall well-being. As many professionals find themselves tethered to their home offices, the lines between personal and professional spaces blur, leading to potential lifestyle changes that can either uplift or undermine physical health. It's crucial to cultivate an environment that encourages regular movement, ergonomic work setups, and structured routines to fend off the detrimental effects of prolonged sitting and screen time. By embedding physical health into our daily remote schedules, we not only nurture our bodies but also invigorate our minds, unlocking the potential for heightened creativity and enhanced resilience. The small, conscious choices we make—from a quick stretch to a mindful walk—can transform our workday, turning the remote experience into one of vitality and balance. In this journey, let's embrace an active lifestyle as an essential part of navigating the complexities of working from home, ensuring that we emerge not just professionally enriched, but physically rejuvenated.

Programs for Holistic Wellbeing are more essential now than ever as our workspaces gravitate towards a virtual existence. In this landscape, organisations are tasked with rethinking their approach to physical health, ensuring that their workforce remains thriving even from afar. Traditional wellness initiatives like gym memberships and in-office yoga classes don't translate directly to remote environments, necessitating a pivot in strategy—anchored on accessibility and adaptability. This shift involves not just maintaining physical health, but embracing a more rounded concept of wellbeing—holistic wellbeing—in the remote work era.

We often underestimate how interconnected our mental, emotional, and physical health are. A holistic approach recognises this interdependence, emphasising programs that cater to more than just the physical aspect. For organisations aiming to prioritise employee wellbeing, offering programmes that encompass lifestyle aspects such as stress management and nutrition will yield more sustainable results. A well-crafted programme doesn't just alleviate physical discomfort but enhances overall satisfaction and productivity.

One practical avenue for such a program is virtual fitness. With the abundance of online platforms, employees can easily access workouts tailored to their needs, whether it's a rejuvenating morning yoga session or a more intense evening boot camp. The benefit of virtual fitness is its flexibility—employees can integrate physical activity into their schedules at their convenience, a boon that traditional office-bound jobs often couldn't offer. Furthermore, these platforms allow for a personalised fitness journey, accommodating various fitness levels and personal goals.

Nutrition also plays a critical role in holistic wellbeing. Offering resources like virtual nutrition seminars and healthy cooking workshops can empower employees to make better dietary choices. As many remote workers are now responsible for all their meals, this type of support can have a meaningful impact on energy levels and overall health. Access to nutritionists for personalised meal planning or dietary advice can further enhance these offerings, ensuring employees have the guidance to make informed choices at home.

Mindfulness and stress reduction techniques are equally vital. In a remote setting, the lines between work and home life can blur, leading to increased stress. Programmes focused on mindfulness practices such as meditation, breathing exercises, and guided visualisations can help create mental space and foster resilience. Organisations can incorporate mindfulness apps or provide subscriptions to platforms that specialise in stress reduction techniques. Encouraging regular breaks and moments of reflection can significantly enhance mental clarity and emotional stability.

Physical health initiatives alone might not be sufficient; integrating emotional and social support systems can bolster holistic wellbeing. Virtual social gatherings, mentorship programs, and peer support groups create a sense of community and belonging, combating the isolation that remote work can often entail. When employees feel connected, they are more likely to be engaged and productive—a critical component of a thriving remote workforce.

Lastly, ergonomics should not be overlooked in a remote setup. Companies should guide their employees on creating ergonomic home offices, ensuring that workspace setups minimise physical strain. Programs offering virtual assessments of home setups or stipends for ergonomic equipment can be particularly beneficial. An ergonomically sound environment not only prevents musculoskeletal issues but also enhances focus and efficiency.

In crafting these holistic wellbeing programmes, organisations should seek input from their employees, tailoring initiatives to the unique needs and challenges of their workforce. By investing in employees' holistic health, companies aren't just enhancing the quality of life for their workers but also laying the groundwork for heightened productivity and long-term success. As we navigate this era, recognising and responding to the nuanced dimensions of wellbeing will distinguish truly forward-thinking organisations from the rest.

Chapter 17:
Economic Implications
of Remote Work

The seismic shift towards remote work is upending traditional economic structures, ushering in both opportunity and upheaval across global markets. As companies embrace this change, they're re-evaluating real estate investments, slashing overheads, and reimagining talent acquisition beyond geographical constraints. Simultaneously, the shift is stitching together a more integrated global workforce while challenging local economies reliant on commuting and office-centric industries. Businesses that lean into this transformation can harness a wider array of skills and drive innovation with unprecedented agility. Yet, it's crucial to acknowledge the rippling effects on urban infrastructure, service sectors, and the fiscal policies of nations. The remote work revolution is more than a logistical pivot; it's an economic evolution beckoning industries to adapt, innovate, and thrive. Embracing these changes with foresight and flexibility can transform potential disruptions into pathways for sustained success.

Impact on Global Economies

Remote work has unleashed a seismic shift in global economies, reshaping the landscapes of traditional employment and national economic strategies. As companies disperse geographically, the talent pool has expanded beyond borders, enabling a more competitive market where skills, rather than location, hold the key to opportunity. This transition enlivens local economies previously untouched by global commerce, offering fresh prospects for innovation and growth. Meanwhile, urban centres experience economic recalibration as demand for office spaces declines, prompting a transformation in real estate and associated industries. Service sectors adapt quickly, creating a ripple effect that challenges long-standing economic models. Balancing these dynamics isn't without its hurdles, but the potential for economic revitalisation and diversification is compelling. As we navigate this evolving economic terrain, the influence of remote work underscores a new era where flexible employment structures shape not just business but the entire world economy.

Predictions for Economic Shifts As remote work continues to redefine how we conduct business globally, the implications for national and global economies are both vast and complex. The shift to remote work doesn't represent just a temporary change but potentially signifies a profound transformation in economic structures and functions. This metamorphosis is expected to bring about various economic shifts, with potential benefits and challenges intricately intertwined.

One pivotal economic prediction is the decentralisation of economic hubs. Historically, major cities have acted as financial and cultural nerve centres, drawing talent and capital to dense central areas. Remote work, however, is breaking this pattern, allowing both employees and companies to operate outside these traditional epicentres. We might see a gradual redistribution of talent and resources, where smaller towns and cities could gain prominence, sparking regional economic rejuvenation. This could herald a renaissance for areas previously overshadowed by sprawling urban landscapes.

The Remote Renaissance

The real estate market is already experiencing shifts in demand due to remote work, but its economic ripples extend further. Corporate investments in high-priced city office spaces might dwindle, reshaping real estate dynamics in urban centres. Conversely, there could be a surge in demand for residential properties in suburban or rural zones. This relocation can spur economic growth in these regions, transforming them into flourishing economic contributors with new businesses and services sprouting to cater to an increased population.

Moreover, the focus on digital infrastructure will intensify. Regions previously considered economically nascent can potentially leapfrog to global prominence if they invest wisely in technology and connectivity. Governments might allocate more resources to enhance digital infrastructure across their territories, fostering innovation hubs in unexpected places. This global race for digital preparedness could create new economic powerhouses, shifting the balance of economic influence worldwide.

With the workforce distributed globally, businesses might harness diverse perspectives to foster innovation. This new economic world order, less constrained by geographic location, could lead to increased competitiveness and collaboration across borders. Industries will feel impelled to adapt quickly, leveraging this diverse talent pool for creative problem-solving and accelerating the pace of innovation, ultimately affecting economic prosperity.

However, these transformations won't come without challenges. Local economies may face periods of instability as they adjust to these shifts. Urban centres, accustomed to bustling economic activities, might struggle with declining demand for services tailored to the traditional office-based workforce. Redundant commercial spaces could result in potential financial losses for real estate sectors unless adaptive reuse and innovative repurposing strategies are implemented.

Additionally, economic disparities between digitally developed and underdeveloped regions may widen without proactive measures. Remote work favours locations with robust digital infrastructures, and regions unprepared for remote integration might lag in economic growth. Therefore, bridging this digital divide becomes essential both to ensure equitable economic opportunities and to harness the benefits of this remote workforce evolution.

An often overlooked aspect is the regulatory adjustments required as economies transition to accommodate remote work. Tax systems, labour laws, and international business regulations will need modifications to reflect the new reality of decentralised workforces. Policymakers will be challenged to create frameworks that support this modern economic landscape while ensuring fair taxation and labour rights protection.

One likely economic shift will involve industry-specific transformations. Some sectors might flourish with remote work, enjoying enhanced efficiency and cost reductions, while others, such as hospitality and transportation, might face adjustments due to diminished daily commutes and business travel. This imbalance requires strategic planning and the development of new business models to sustain industry viability in a remote-centric future.

The global economy may see fluctuating productivity patterns as businesses adapt to the remote mode of operation. While remote work can boost productivity through flexible working conditions and reduced commuting stress, it can simultaneously pose challenges in maintaining workforce collaboration and cohesion. Navigating these dynamics will require innovative management strategies and continuous adaptation for economic growth to be sustainable.

Another potential economic implication concerns the shifting dynamics in workforce expenditure. Employees working remotely are likely to invest differently, which can alter consumer behaviour.

Spending may shift towards home office equipment, digital tools, and home maintenance, reducing expenditure on commuting, office attire, and in-office dining. The ripple effect of these changes might prompt businesses in affected sectors to reinvent their offerings, driving economic reinvigoration.

Finally, the global remittance economy could see an increase. With people no longer bound by geographical work constraints, there's a potential rise in expatriates or transnational workers who send earnings back to their home countries. These remittances can bolster local economies, providing vital injections of cash flow that support development initiatives, increase consumption, and facilitate economic growth in those regions.

In conclusion, while there lies considerable potential for economic growth and innovation through the remote work model, it's essential to approach these predictions with both optimism and strategic foresight. By anticipating and preparing for these economic shifts, nations and businesses can harness the full potential of remote work, driving prosperity and inclusivity in the global economic landscape. The future holds promise, but realising that promise requires action, insight, and a willingness to evolve with the changing tides.

Chapter 18:
The Future of Remote Education

The future of remote education is poised to revolutionise the learning landscape, integrating cutting-edge technology with flexible approaches to make education more accessible than ever. As digital learning environments evolve, they promise to personalise education, making it possible to tailor learning experiences to an individual's strengths and weaknesses. This shift doesn't merely aim to supplement traditional forms of education but seeks to redefine them, accommodating diverse learning styles and breaking geographical barriers. Imagine a world where students from varied backgrounds connect and collaborate as part of a global classroom, enabled by innovative platforms that make knowledge exchange as seamless as turning a page. As we stand on the brink of this educational metamorphosis, it is crucial for professionals and educators alike to harness the opportunities presented by this digital era. By fostering a culture of continuous learning and adaptation, we can prepare for the challenges and possibilities that will shape tomorrow's educational achievements, ultimately driving both personal and societal progress.

Shaping the Landscape of Learning

As remote education continues to evolve, it's transforming how we perceive learning and integrating it within our daily lives. The virtual classroom isn't just a temporary fix; it's becoming a cornerstone of modern education systems, breaking geographical barriers and accessing a wealth of global resources. Professionals navigating remote work will find the merging of technology and educational methodologies pivotal in honing skills and fostering adaptability. By embracing this shift, HR leaders and corporate managers can promote lifelong learning, ensuring that their teams stay at the forefront of industry developments. This dynamic educational landscape encourages a culture of curiosity and growth, motivating individuals to seize every opportunity for personal and professional development. The emphasis is on nurturing not just technical competencies but also enhancing critical thinking and creativity, ultimately shaping a workforce that's prepared for the challenges and opportunities of a remote-first world.

Trends in Online Education are reshaping the landscape of learning, marking a profound shift in how knowledge is accessed and disseminated. As we delve into the future of remote education, it's crucial to acknowledge the transformative role that online education plays. In recent years, traditional learning environments have given way to flexible, accessible, and highly personalised educational experiences. This trend is not just reshaping classrooms but reimagining them, offering students and professionals alike the opportunity to learn without the constraints of geography.

The demand for online education has surged, fuelled by the growing need for lifelong learning and re-skilling. Professionals constantly seek to enhance their skills and knowledge to stay relevant in rapidly changing industries. Consequently, institutions and educators are tailoring their offerings to appeal to a diverse, global audience. Micro-credentials, online certifications, and modular courses provide a customised learning journey, empowering individuals to chart their own educational paths.

Technological advancements are driving these trends forward at an unprecedented pace. With artificial intelligence and machine learning making strides, educational platforms are becoming more intuitive and adaptive. These technologies personalise the learning experience, assessing each learner's strengths and weaknesses, and recommending resources that suit their individual needs. This approach fosters a more engaging and effective learning environment, as learners receive content that resonates with their unique needs and aspirations.

Moreover, the integration of virtual and augmented reality in online education is creating immersive learning experiences that were once unimaginable. By simulating real-world scenarios, these technologies enhance understanding and retention, making complex concepts easier to grasp. For example, medical students can practise surgery in a virtual environment, while engineers can design and test prototypes without the constraints of physical resources. Such innovations hold the promise of levelling the educational playing field, providing all learners with access to cutting-edge resources.

However, it's not just the technology that is evolving. Pedagogical approaches are adapting to this new educational landscape. The emphasis is shifting from passively receiving information to actively constructing knowledge. Collaborative learning tools, forums, and digital classrooms facilitate interaction and engagement among peers, making learning a communal rather than solitary activity. Such methods encourage critical thinking and creativity, as learners engage in discussions and debates that challenge their viewpoints.

Another significant trend is the growing popularity of asynchronous learning. This flexible approach allows learners to access course materials at their convenience, breaking down the barriers of time and location. Whether one's juggling a demanding job or personal commitments, asynchronous learning offers the flexibility needed to pursue educational goals without sacrificing other responsibilities. This adaptability is particularly vital for professionals seeking to advance their careers while maintaining a work-life balance.

The Remote Renaissance

Despite the myriad benefits, online education is not without its challenges. One of the most pressing issues is ensuring equitable access to technology and the internet. Disparities in digital access can exacerbate educational inequalities, underscoring the need for policies and initiatives that strive to make technology accessible to all. Moreover, ensuring that online education maintains quality standards equivalent to traditional classroom learning is an ongoing endeavour that requires cooperation between educators, institutions, and policymakers.

Furthermore, student engagement remains a critical concern. While technology can facilitate convenient access to education, it can also foster a sense of isolation if not navigated thoughtfully. Hence, building a sense of community online is paramount. Introducing gamification elements, encouraging peer-to-peer interaction, and utilising social media platforms are some strategies that institutions are using to cultivate engaging learning environments.

A noteworthy trend is the increasing collaboration between educational institutions and industry partners. By aligning curriculum with industry needs, learners can gain relevant skills that are immediately applicable to the workforce. Internships, project-based learning, and industry-led workshops equip learners with the practical know-how to excel in their chosen fields. Such partnerships bridge the gap between academia and industry, ensuring that education remains relevant and responsive to market demands.

Moreover, the global nature of online education is fostering cultural exchange and understanding. Learners from diverse backgrounds and geographies can interact and collaborate on projects, gaining insights into different perspectives and practices. This cross-cultural engagement is enriching the learning experience and preparing students for a globalised workforce where adaptability and cultural sensitivity are prized skills.

As we look to the future, online education promises to continue evolving in response to societal, economic, and technological trends. As digital natives become digital leaders, their expectations will shape the future of learning. Educational institutions must be agile, continually innovating and adapting to meet the needs of these learners who have grown up in an era defined by rapid technological change.

In conclusion, the trends in online education reflect a broader shift towards a more inclusive, flexible, and learner-centric approach. By embracing these trends, we can shape the future of remote education into a powerful tool for personal and professional growth. It's a future where learning is no longer constrained by boundaries, but enriched by them, creating a landscape of endless possibilities and opportunities.

Chapter 19:
Remote Work Across Industries

In the tapestry of modern work, industries have woven unique threads of remote work methodologies, spotlighting the diversity of adaptation across sectors. From tech giants pioneering digital nomadism to healthcare professionals embracing telemedicine, the shift reshapes not only where but how we work. Traditional boundaries blur as manufacturing outfits harness remote monitoring technologies while education reinvents pedagogy with online classrooms. Yet, these shifts aren't without their hurdles; each industry faces niche challenges—security concerns in finance, collaboration complexities in creative fields. However, with every hurdle comes opportunity. Industries are innovating tailored solutions, catalysing growth and endurance in remote settings. As professionals, adapting means more than just embracing change; it demands a pioneering spirit to redefine norms and carve new paths in an ever-evolving work landscape.

Sector Specific Adaptations

As remote work continues to reshape how industries operate, sector-specific adaptations reveal a fascinating tapestry of innovation merging traditional models with new technologies. In sectors like healthcare, telemedicine thrives as it breaks geographical barriers, while retail navigates the digital transformation through e-commerce and virtual customer engagement. The tech industry, naturally equipped for remote work, evolves further to enhance collaborative tools and cyber security measures. Meanwhile, in education, remote methodologies revamp teaching strategies, allowing for a seamless experience across diverse platforms. Each sector adapts in its unique way, weaving flexibility and resilience into their operational fabric, showcasing a remarkable ability to not just survive, but thrive in this new era. This persistent drive to adapt underscores an inspiring truth: when faced with challenge, innovation doesn't just respond—it flourishes, paving the way for a future where work knows no boundaries.

Unique Challenges and Solutions Navigating remote work across various industries presents unique challenges, each demanding innovative solutions to maintain efficiency and connectivity. As sectors adapt, the nuances of their needs become pronounced, presenting obstacles that are as diverse as the industries themselves. Yet, with these challenges come unparalleled opportunities to redefine how work is approached, ensuring not only continuity but also growth in a digital age.

In the world of healthcare, for instance, the very essence of work is inherently hands-on, creating a paradox in the realm of remote work. Health practitioners face the daunting task of ensuring patient care through telemedicine, an adaptation that requires robust digital infrastructures and sensitive handling of personal data. Clinics and hospitals are deploying solutions like virtual consultations, harnessing AI-driven diagnostic tools, and investing in secure data storage systems. The challenge lies in marrying technology with patient confidentiality and care, ensuring that remote doesn't mean removed.

Then there's the education sector, witnessing a tectonic shift in both teaching and learning modalities. The pivot to virtual classrooms has been swift yet fraught with technological disparities. Educators wrestle with delivering engaging content without physical presence, relying on digital platforms that need to be both intuitive and inclusive. Creative solutions like gamified learning experiences and virtual labs are bridging the gap, addressing both the distraction problems of home environments and the need for interactive learning. However, the digital divide remains a formidable barrier, pointedly requiring initiatives for equal access to technology and internet connectivity.

Financial services have their own set of intricacies. As customer interactions go digital, banks and financial institutions must adapt to maintain both accessibility and trust. The challenge of ensuring data security while offering seamless customer service is significant. Implementing solutions like biometric authentication, blockchain for secure transactions, and enhanced digital customer service tools help mitigate risks. Moreover, training employees to handle complex financial situations remotely requires a parallel focus on skill development through e-learning modules tailored to the sector's ever-evolving needs.

The creative industries face a unique set of challenges, primarily centred around collaboration. Artists, designers, and writers thrive in environments of spontaneous interaction and ideation, which are often constrained in remote settings. To tackle this, many creative teams have leveraged advanced collaboration tools that allow for real-time co-creation and feedback. These tools, combined with virtual brainstorming sessions, attempt to replicate the studio environment online, although the debate rages on about whether digital can truly match the creativity sparked through in-person encounters.

Manufacturing and logistics sectors, traditionally grounded in physical operations, have embraced the challenge with innovative technological adaptations. Remote monitoring systems and IoT solutions are now commonplace, allowing managers to oversee production lines from afar. The implementation of digital twins—virtual replicas of physical entities—enables simulation and optimisation of processes remotely, a development that has markedly reduced downtime. The challenge remains in upskilling workers to operate advanced technologies, necessitating a shift in workforce training approaches.

Legal practices, often reliant on face-to-face consultations and in-person court proceedings, find themselves adapting through video conferencing and digital document management systems. The transition has brought about challenges around maintaining client confidentiality and managing time zones for international cases. Solutions lie in adopting encrypted communication platforms and flexible scheduling practices that respect the global nature of legal work today.

For technology companies themselves, the shift to remote work has arguably been smoother, yet not without challenges. Innovating in isolation can stall momentum, and maintaining team synergy becomes crucial. Companies are experimenting with virtual hackathons and collaborative coding sessions to keep team spirit alive and continuous integration practices smooth. Furthermore, maintaining a company culture that embodies creativity and innovation can become muted when interactions are screen-bound. A solution has been in structured online mentorship programmes that promote cross-team learning and continuity of corporate ethos.

In the realm of retail, the symbiosis between physical and digital spaces has seen rapid evolution. E-commerce platforms have thrived, yet the challenge rests in creating a seamless customer experience that aligns with the brand's physical presence. Retailers are investing in AR and VR technologies to enhance online shopping with immersive experiences, offering solutions that bridge the sensory gap in virtual stores.

Finally, the public sector encounters its unique set of hurdles. Government services transitioning to a remote setup need to ensure transparency and accessibility to citizens, which is no easy feat. Solutions involve leveraging robust e-governance platforms that enable service delivery with accountability. Cybersecurity remains a persistent concern, prompting ongoing investments in advanced threat detection and response systems.

Across these diverse arenas, the key takeaway is the necessity for adaptive leadership and creative problem-solving. As remote work continues to evolve, so must the strategies that support it, ensuring that each sector not only meets its specific challenges head-on but also thrives in the digital age. These solutions underscore the capacity for innovation inherent in human enterprises, reminding us that while the challenges are unique, the potential for ingenuity is limitless.

Chapter 20:
Crisis Management in
a Remote World

Navigating a crisis in a remote work environment demands a paradigm shift in traditional management approaches. This chapter underscores the necessity of foreseeing potential disturbances and equipping teams with robust, flexible strategies that transcend geographical boundaries. In a world where face-to-face reassurance isn't an option, building trust and clear lines of communication become paramount, ensuring that each team member understands their role in a crisis. By embracing digital tools tailored for emergency response and fostering a culture of resilience, leaders can instil confidence even in the most turbulent times. Yet, technology alone doesn't suffice. It requires a marriage of empathy, agility, and foresight—qualities that empower remote teams to not just survive but thrive amid uncertainty. Adapting swiftly to these new methods isn't just about maintaining operations; it's about redefining them for a future where the unexpected becomes the new normal. This transformation is as much about mindset as it is about logistics, urging organisations to innovate and regroup with courage when facing the next big challenge.

Handling Emergencies Remotely

In an era where face-to-face intervention isn't always feasible, handling emergencies remotely requires both agility and foresight. The key is integrating technology and human touch, empowering individuals and teams to act swiftly yet thoughtfully. It's about building an infrastructure of trust and reliability through clear protocols and technologies that can connect people instantly and securely. When emergencies strike, the preparedness to pivot and communicate becomes essential, ensuring solutions are just a click away. Teams need to be equipped not just with the right tools, but with confidence and knowledge that their well-coordinated efforts can surmount crises from afar. As we master these elements, we're not just safeguarding business continuity; we're catalysing a resilient and responsive remote work culture.

Planning and Response Strategies have become pivotal in the era of remote work, especially when handling emergencies remotely. Imagine a sudden cyber-attack threat that could jeopardise a company's sensitive data. Or think about a natural disaster that disrupts the communication network of a geographically scattered team. Such scenarios demand rapid and adept responses, meticulously planned in advance. It's more than crisis management; it's about ensuring an organisation can withstand and navigate through uncertainties, coming out more resilient on the other side.

The first step in crafting effective planning and response strategies is risk assessment. Understanding the complexity and variety of potential emergencies in a remote setting is crucial. Face-to-face scenarios often come with immediate visual and contextual cues that can guide responses. However, in a remote work environment, these signals may not be as readily apparent. Therefore, organisations must delve into digital risk landscapes, identifying vulnerabilities that may not be obvious at first glance. This involves an audit of technological infrastructure and data security policies, which serve as the backbone for remote operations.

A fundamental aspect of planning involves clear communication protocols. In the event of a crisis, clarity can be a literal lifeline. Organisations should establish clear guidelines on who should be informed, how quickly information needs to be disseminated, and through what channels. The use of digital communication tools must be optimised to serve not just routine operations but also crisis scenarios. Regular practice drills and simulations can help teams become familiar with these protocols, making them more effective when emergencies actually occur.

One can't overlook the importance of leadership in these strategies. Remote crisis management requires leaders who are not only decisive but also empathetic. They need to balance providing direction with understanding the challenges their teams face, perhaps thousands of miles apart and under duress. Training leaders to operate in this dual capacity is critical. Workshops and role-playing scenarios can serve as effective methods to prepare leaders for handling emergencies remotely, ensuring they maintain team cohesion and morale when it's most needed.

Moreover, companies need to build a culture of resilience. This means fostering an environment where employees are encouraged to voice concerns and suggest solutions, creating a more responsive and agile organisational system. In doing so, employees themselves become part of the crisis management process. They're not just passive recipients of decisions but active participants in the business continuity plan. This fosters a sense of ownership and accountability, critical in any emergency response scenario.

Another crucial consideration is the technology stack that supports remote work. As companies increasingly depend on digital platforms for everyday operations, these tools must be chosen with crisis management in mind. Redundancy becomes key—having backup systems and alternative communication channels ensures that a single point of failure doesn't cripple the entire operation. Cyber incidents or downtime in one tool should not halt productivity or emergency communication.

It's essential to drive home the idea of collaboration in crisis strategy. While each department may have specialised needs during an emergency, the overall organisation must work as a cohesive unit. Cross-functional teams are instrumental in unifying different aspects of the business, ensuring a seamless response. Regular meetings between department heads to anticipate potential disruptions and outline coordinated responses can make all the difference when speed and efficiency are vital.

Flexibility remains a cornerstone of any planning and response strategy. The landscape of remote work evolves continuously, and so do the challenges. A stagnant plan soon becomes outdated. Organisations must commit to regularly updating their strategies based on the latest technological advances and emerging threats. This constant state of evaluation and adaptation allows for refined processes that better serve the organisation's remote workforce.

Additionally, psychological readiness is as important as logistical preparedness. Employees working remotely might experience crises differently, with potential isolation exacerbating stress. Providing mental health support should be integrated into emergency response plans, ensuring that along with addressing the immediate crisis, the wellbeing of employees is not neglected. Access to virtual counselling services or a dedicated helpline can ease the burden staff might feel in overwhelming situations.

Education and training are indispensable regarding the success of planning and response strategies. Regular workshops, accessible webinars, and interactive e-learning modules can empower employees with the necessary skills and knowledge to handle emergencies effectively. This preparation goes beyond just technical know-how and includes developing soft skills such as adaptability and clear communication under pressure.

Furthermore, feedback loops should be established to refine plans continuously. Post-crisis evaluations, involving stakeholders to assess strategy efficacy, can provide valuable insights. By analysing what worked and what didn't, organisations can iterate on their plans, making them more robust and tailored to their unique needs. A transparent feedback mechanism ensures that learning from past experiences is a shared practice, further enriching the company's strategic preparedness.

Planning and response strategies in a remote world call for a blend of foresight and flexibility. By weaving together comprehensive risk assessment, resilient leadership, effective communication, and an adaptive mindset, organisations can not only navigate the challenges of remote emergencies but thrive amidst them. These strategies reflect a dynamic approach that underlines the importance of preparedness and the invaluable human element within remote work environments. Emphasising proactive planning helps cultivate a culture that can overcome any crisis, turning potential vulnerabilities into a testament to organisational strength.

Chapter 21:
Remote Work Policy Development

Designing remote work policies is a crucial task that demands both foresight and adaptability. It requires us to balance organisational goals with employee needs, creating frameworks that foster productivity while ensuring flexibility. The challenge lies in engineering policies that accommodate diverse roles and functions, and are resilient to evolving work paradigms. We must consider factors like communication protocols, data security, and performance metrics, embedding them into the fabric of our policies. Crafting these policies isn't just about setting rules; it's about shaping a future-oriented workplace culture that embraces change and catalyses growth. By engaging all stakeholders in this process, we lay the groundwork for a workforce that is agile, inclusive, and innovative, meeting the complexities of remote work head-on. This isn't merely a strategic move; it's an inspiring call to transform the traditional work landscape into one that thrives on trust and empowerment. Let these policies not only guide our actions but also inspire a collective vision that propels us forward into a new era of work.

Crafting Effective Policies

Crafting effective remote work policies is an exercise in foresight, compassion, and adaptability. In the dynamic landscape of modern work, these policies serve as a framework that balances organisational objectives with the diverse needs of a distributed workforce. To design policies that are both practical and inspiring, it's essential to cultivate a deep understanding of remote work's nuances and challenges. This means acknowledging the need for flexibility while ensuring accountability and productivity. Leaders must embrace an employee-centred approach, fostering trust and inclusivity, and leveraging technology to enhance communication and collaboration. Effective policies aren't just about rules—they're about creating environments where remote workers feel valued and empowered, reflecting a commitment to both organisational success and individual well-being.

Guidelines for Implementation Developing effective remote work policies is a journey that involves strategic insight, understanding of workforce dynamics, and a keen eye for adaptability. The transition to remote work isn't just a change in location; it's a transformation in how organisations communicate, motivate, and measure success. As we embark on this process, it's crucial to first acknowledge the diversity of work styles and the plethora of tools available. A one-size-fits-all policy can lead to organisational discord, so tailoring strategies with flexibility at their core is essential.

To begin with, the clarity of purpose within remote work policies can't be overstated. Organisations must delineate the objectives these policies aim to achieve. Are they primarily to increase productivity, enhance employee satisfaction, or ensure continuity of operations? By pinpointing these goals, it's possible to craft policies that are not only comprehensive but also relevant. Additionally, involving a broad spectrum of stakeholders in the policy-making process is instrumental. Including voices from different departments, as well as from various hierarchical levels, ensures the policies are inclusive and cater to the varying needs across the board.

Next, consider the role of technology as the backbone of remote work. Selecting the right tools and platforms to facilitate communication, collaboration, and project management is crucial. It's not merely about adopting the latest technologies but understanding the unique needs of your team and matching those needs with the right solutions. Pilot testing new tools with a smaller group before organisation-wide implementation can reduce potential resistance and technical hiccups.

Another critical guideline is maintaining open channels of communication. Regular check-ins and the use of virtual platforms to conduct meetings and brainstorming sessions can foster connectivity, even across distances. Establishing these routines not only ensures alignment with organisational goals but also mitigates the feeling of isolation often associated with remote work. Encouraging feedback from employees regarding the policy's effectiveness and how it affects their daily work life helps in crafting a more dynamic policy that evolves with the team's needs.

In effectively implementing remote work policies, flexibility should be a guiding principle. Flexibility in work schedules, task management, and even in evaluating performance can make a world of difference. The traditional 9-to-5 model may need to be re-evaluated in favour of more results-oriented metrics. What counts now is not the number of hours spent in front of a screen but the outcomes delivered, thereby empowering employees to tailor their workday in a manner that maximises productivity.

Moreover, consider incorporating guidelines on cybersecurity and data privacy. With employees accessing company data from various locations and networks, understanding and mitigating risks becomes a priority. Training sessions on safe internet practices and regular audits of security protocols are necessary measures to protect both the employee and the organisation. A policy that anticipates potential security challenges while providing clear guidance on data handling is non-negotiable.

The human aspect should never be overlooked. A remote work policy that accounts for mental health, work-life balance, and a support system for employees will have a more profound and lasting impact. Providing resources such as mental health days, access to counselling services, and promoting wellness programs can significantly enhance employee well-being and engagement. Recognising the unique challenges posed by remote work, such as social isolation or burnout, is fundamental in creating a supportive environment.

Lastly, periodically reassessing and updating the policies ensures they stay relevant and effective in a landscape that is continuously evolving. This adaptability supports a culture of continuous improvement and ensures that the organisation remains agile in the face of new challenges. Regular reviews of the policy's impact on the workforce and its alignment with organisational objectives should be standard practice.

As we navigate through these guidelines for implementation, the ultimate goal remains clear: to create a work environment that is as effective and humane from a distance as it is in a traditional office setting. Ensuring that remote work policies are thoughtfully crafted, effectively implemented, and continuously refined will lay a strong foundation for achieving this goal.

Chapter 22:
Sustainability and the
Remote Workforce

As we reach the juncture where remote work intertwines with sustainability, it's clear that the modern workforce stands at the forefront of an environmental revolution. By cutting down daily commutes, remote work significantly reduces carbon footprints and inspires a shift towards greener practices. Individuals now have the choice to adopt more sustainable lifestyles without the constraints of city-bound living. Companies, too, are rethinking their environmental responsibilities, opting for virtual solutions that demand less from our planet's resources. Yet, this transformation is not without its challenges. Issues like increased home energy consumption and the digital carbon footprint must be addressed to truly harness the sustainability potential of remote work. To this end, fostering an ethos of environmental awareness within corporate policies not only aligns with the broader goal of sustainability but also galvanises a workforce committed to creating positive change. As we navigate this path, the collective action of individuals and organisations alike will determine our ability to champion a sustainable future in a rapidly evolving remote work landscape.

Environmental Benefits and Challenges

The shift to a remote workforce isn't just transforming the way we work, it's also reshaping our environmental footprint. As commuting fades into a relic of the past, we see a direct reduction in carbon emissions and a decrease in the daily grind of urban traffic congestion. The potential for remote work to foster a more sustainable world is immense, yet we mustn't overlook the challenges it presents. Increased reliance on technology means energy consumption and electronic waste are on the rise, calling for smarter solutions and greener practices. Balancing these ecological gains with the hurdles requires bold action and collective ingenuity. It's not solely about reducing carbon footprints; it's about reimagining a future where the workforce is as environmentally conscious as it is productive, ensuring that the shift benefits our planet just as it benefits businesses and individuals.

Promoting Sustainable Practices extends beyond merely recognising the environmental benefits and challenges of remote work; it calls us to engage actively with change. As remote work gains traction, our obligation is not just to adopt these practices but to wield them as tools for broader ecological impact. Remote work, in its essence, offers a unique opportunity to draft a blueprint for sustainable living—one where conscientious policies and actions lead to significant environmental benefits. But how do we ensure that these opportunities translate into real-world positive outcomes? By building on the inherent benefits and addressing the pressing challenges, we can cultivate a more sustainable remote workforce.

The reduction in commuting is often highlighted as a clear environmental benefit of remote work. Fewer cars on the road mean reduced emissions. But, promoting sustainable practices within remote work isn't solely about cutting down on travel. It also involves the adoption of more efficient energy usage within home offices. This transition leads us to consider the characteristics of remote workspaces:

are they powered by renewable energy? Do they use energy-efficient devices? By focusing on these aspects, companies can incentivise employees to opt for greener alternatives—perhaps offering subsidies for solar panels or energy-efficient appliances.

Implementing practices that minimise our carbon footprint also includes reducing waste generated during work processes. A significant proportion of traditional office waste originates from excess paper usage. Remote work offers the perfect backdrop for a paperless office. Encouraging digital documentation and cloud storage solutions reduces clutter and the environmental toll of paper production and waste. Such a shift demands investment in secure digital tools and platforms—allowing for efficient document handling without compromising data security.

It's not just about what we do in our compact home offices but how we connect. Virtual meetings, a staple in remote work, consume significantly less energy compared to large, in-person conferences that require travel, venue arrangements, and other resources. Businesses should leverage these tools not just to connect teams seamlessly but to further drive a culture of minimalism and efficiency in communication.

However, the conversation around remote work and environmental sustainability cannot ignore the potential oversights. Increased reliance on digital infrastructure can lead to higher energy demands; data centres—the backbone supporting our cloud-based workflows—consume vast amounts of power. Here lies a challenge: optimising digital operations to balance or negate this increased consumption. Encouraging the use of platforms and providers committed to green data centres is a step towards addressing this less visible but crucial aspect.

Moreover, the sustainability conversation should not be divorced from the well-being of remote employees themselves. A sustainable practice considers both environmental and human aspects. Providing guidance on setting up ergonomic and well-ventilated home offices is vital. It ensures not only health and comfort but also energy efficiency—leading to healthier employees and responsible energy use.

Businesses should also consider the potential challenges of decentralised work routines which might blur the lines separating work and personal time. The constant connectivity can lead to burnout, an issue that impedes sustainable human practices. Companies can mitigate this by advocating for mindful work habits and promoting digital wellness. This includes fostering an environment where employees feel enabled to disconnect outside of work hours to recharge and engage in healthier lifestyles, benefiting their own well-being and the environment.

Sustainable remote work practices must also align with broader corporate social responsibility objectives. When companies publicly commit to environmental sustainability, it embeds value-driven goals into the very fabric of their operational culture. Remote work provides the ideal setting to sponsor team challenges that promote sustainability, such as reducing plastic usage or participating in local community greening projects—all of which contribute to longer-term environmental impacts.

Beyond immediate benefits, promoting sustainable practices in remote work involves a commitment to continual learning and innovation. Organisations should consider investing in training programs focused on environmental awareness and sustainable practices. Empowering employees with knowledge and tools to make environmentally conscious decisions can ripple outwards, influencing their perceptions and actions both within and beyond their professional lives.

Lastly, collaboration between companies in the remote work ecosystem can amplify the impact of individual sustainable initiatives.

Creating alliances and networks that exchange best practices can lead to industry-wide improvements, setting a standard for responsible remote work practices globally. Sharing successes and learning from failures allows organisations to collectively strive towards a common goal—one that prioritises the health of our planet and its inhabitants.

As professionals navigating the landscape of remote work, it's essential to grasp the unique chance we have to promote sustainability actively. The challenge isn't in recognising the potential but in implementing practices that align with our values and broader environmental goals. Embracing this change requires a collective effort and creative strategies that transcend habitual practices—urging us forward with hope and a commitment to a more sustainable future.

Chapter 23:
Virtual Collaboration
on a Global Scale

In the grand tapestry of remote work, virtual collaboration on a global scale stands out as one of the defining patterns of this era. The ability to bridge teams across international borders isn't just a testament to technological advancement; it's a powerful narrative of human adaptability and ingenuity. As organisations dive into this global pool, they encounter the vivid spectrum of cultural diversity and the often-challenging dance of synchronising multiple time zones. Yet, the rewards of such endeavours are immense. By harnessing varied perspectives and diverse skills, companies not only foster innovation but also cultivate a richer corporate culture that transcends geographical limitations. Embracing this mode of collaboration demands more than just effective communication tools; it requires a willingness to develop empathy and cultural intelligence. Organisations that achieve this manoeuvre the delicate balance between global reach and local nuance, ultimately crafting a workforce that is as cohesive as it is diverse. It's an inspiring moment in the professional world, where the shared goal of innovation through collaboration becomes a true catalyst for progress.

Bridging International Teams

In the complex tapestry of global collaboration, bridging international teams emerges as both a challenge and an opportunity. As remote work redefines geographical boundaries, it invites us to weave a richer, more diverse workforce that thrives not despite differences, but because of them. The key is crafting an environment where cultures intermingle, creating a symphony of ideas that resonate more powerfully together than apart. This requires intentionality, a focus on building trust, and a commitment to understanding varied communication styles. By leveraging technology and fostering a culture of inclusivity, businesses can cultivate teams that are not only geographically diverse but also globally effective. In doing so, they can create workplaces that embody resilience and innovation, becoming beacons of adaptability in an interconnected world. Harnessing the potential of diverse perspectives and experiences, organisations are better positioned to navigate the complexities of global markets and to inspire growth that transcends borders.

Overcoming Cultural and Time Zone Barriers The rapid shift towards virtual collaboration on a global scale is transforming how we work, but it brings unique challenges. Among the most significant are cultural differences and time zone barriers. In the diverse tapestry of an international team, misunderstandings can stem not only from linguistic differences but also from cultural nuances. The richness of global collaboration is undeniable; however, it requires us to be acutely aware and respectful of the varied perspectives each team member brings to the table.

Cultural differences manifest in various ways, from communication styles to decision-making processes. While some cultures might value directness and brevity, others might appreciate a more nuanced and diplomatic approach. This contrast can affect everything from email exchanges to virtual meetings. For instance, a simple "yes" in one culture could imply agreement and commitment, whilst in another, it might merely signal that the message was received and understood. It's crucial, therefore, to foster an environment where team members feel comfortable clarifying their intentions and expectations.

Understanding cultural contexts can significantly enhance virtual collaboration. Providing cultural awareness training is a proactive way to mitigate misunderstandings. This training could involve learning about colleagues' national holidays, traditions, and typical working hours, ensuring that everyone's cultural identity is recognised and respected. Emphasising cultural sensitivity in a virtual setting not only helps avoid faux pas but also strengthens team cohesion by encouraging an inclusive environment where diverse ideas can flourish.

Time zone barriers, on the other hand, add another layer of complexity to international collaboration. Working with colleagues across different time zones requires a level of flexibility and patience that isn't typically necessary when everyone shares the same working hours. This challenge calls for creative solutions to ensure that the team's productivity doesn't wane and that no team member feels isolated or left behind. One effective approach is to harness asynchronous communication tools, which allow team members to contribute and respond at different times without interrupting each other's workflow.

Schedulers and shared calendars have emerged as indispensable tools in bridging time zone gaps. By clearly marking overlapping hours, these tools help teams identify potential meeting times that work for everyone. Yet, these technologies are only as effective as the people who use them. Leaders in international teams must encourage a culture of transparency with regards to availability while ensuring that workload and expectations are adjusted to consider time zone disparities.

Customary approaches to meetings must also evolve. Rather than hosting lengthy, one-size-fits-all meetings that may inconvenience colleagues in distant time zones, teams can adopt a modular meeting approach. This involves breaking down discussions into smaller, focused sessions that cater to relevant participants only, thereby respecting individual time constraints and maximising collective input.

More than just altering meeting structures, there's a pressing need for fostering trust and accountability in the remote workforce. Without the benefit of face-to-face supervision, team members must rely on mutual trust to assume responsibilities diligently and transparently. By maintaining clear and open lines of communication and regularly checking in with each other, teams can cultivate a sense of mutual respect and understanding, which is vital in overcoming these geographical divides.

Additionally, empathy becomes a fundamental cornerstone in bridging international teams. Being aware of the challenges that colleagues face due to cultural and time zone differences demonstrates consideration and respect, which can significantly boost morale and collaboration. Simple gestures, like sending a follow-up message acknowledging someone's late-night contribution, can reinforce a sense of togetherness and shared purpose.

The importance of empathy and cultural awareness can't be understated, especially when considering the different public holidays and significant events that colleagues across the globe might observe. Being mindful of these can prevent the unintended pressure or awkwardness of requesting work during someone else's festive period. This is where the power of inclusive scheduling and empathetic leadership comes into play, demonstrating a commitment to not only respect but also celebrate cultural diversity within teams.

Another critical factor in bridging global collaboration efforts is language flexibility. While English is often the de facto language for business, many team members might be non-native speakers. It's important to foster an environment where clarity is valued over complexity, with sensitivity towards linguistic challenges that some may face. Encouraging the use of straightforward language and offering translation resources can help in ensuring that each team member feels confident and empowered to share their thoughts without the barrier of language.

Ultimately, overcoming cultural and time zone barriers requires an ongoing commitment to learning and adapting. Teams must be open to evolving their practices and norms as they gain new insights into the preferences and challenges of their colleagues. By embodying a culture of learning, teams ensure that they remain dynamic and resilient in the face of the ever-changing demands of global virtual collaboration.

In conclusion, while cultural and time zone barriers pose significant challenges to global collaboration, they also present opportunities for growth and innovation. By embracing cultural differences as strengths and utilising technology to navigate time zone challenges, teams can forge stronger, more inclusive partnerships. This not only enhances productivity but also enriches the work experience for everyone involved, paving the way for a more connected and empathetic workplace on a global scale.

Chapter 24:
Preparing for the Future of Work

As we stand at the cusp of a dramatic transformation in how we work, it's vital to embrace the future with optimism and adaptability. The pandemic has acted as a powerful catalyst, accelerating trends in remote work and reshaping traditional workplaces. Now, the challenge lies not only in keeping pace with rapid technological advancements but also in fostering a work environment that supports continuous learning and innovation. Organisations must leverage emerging tools poised to redefine workforce dynamics, from AI-driven analytics predicting employee needs to immersive virtual environments enhancing cross-border collaborations. As leaders and professionals, we must prioritise agility, frequently reassessing our strategies to align with shifting expectations and new opportunities. Ultimately, preparing for future work requires a delicate balance between harnessing cutting-edge technologies and nurturing the human elements crucial for connectivity and creativity. The future beckons those willing to innovate, adapt, and lead with courage.

Forecasting Remote Work Trends

As we cast our gaze towards the horizon of work's future, the landscape appears to be teeming with transformation and opportunity. The era of remote work isn't just a temporary shift; it's evolving into a cornerstone of modern employment structure. We can envision a future where flexible workspaces aren't merely a perk but a standard, allowing organisations to tap into a global talent pool without geographical constraints. The nexus of technology and human creativity will continue to push the boundaries of what remote work can accomplish. Industries will drive innovation as they adapt to changing needs, crafting tools and frameworks that amplify productivity and foster deep, genuine connections. In this brave new world, companies must nurture cultures that are resilient and inclusive, interweaving diversity and adaptability as keys to thriving amidst change. The fabric of work is being rewoven, and with it comes the chance to redefine the very essence of our working lives, making it imperative for professionals, leaders, and companies to anticipate and embrace these seismic shifts with foresight and agility.

Innovations on the Horizon As the realm of remote work continues to evolve, it's becoming clear that the next wave of innovations will radically reshape not just how we work, but how we live and interact. These developments promise to enhance productivity, foster collaboration, and ensure that the essence of human connection isn't lost in a virtual world. By proactively embracing these innovations, organisations and individuals can position themselves at the forefront of this ongoing transformation.

Among the most anticipated trends is the rise of virtual reality (VR) and augmented reality (AR) technologies. Imagine stepping into a virtual office where your colleagues appear around you in holographic form, enabling face-to-face interactions irrespective of where you are physically located. This kind of immersive experience could bridge the gap between remote and in-office work, making collaboration and innovation possible on an entirely new level. These tools are not just conceptual ideas; companies are already investing heavily in their development, signalling a future where digital presence might become as impactful as physical presence.

Then there's the potential of artificial intelligence (AI) and machine learning. As remote work mandates a higher degree of self-management, AI-powered tools can assist in personalising work experiences, automating routine tasks, and offering data-driven insights for decision-making. AI assistants could handle scheduling, prioritising tasks, or even managing team communications, freeing up time for human workers to focus on more strategic and creative contributions. This integration of AI into daily workflows may lead to not only heightened efficiency but also more meaningful work engagements.

Beyond tools and systems, the concept of a "workplace" itself is set for a transformation. Next-gen workspaces will likely be increasingly flexible and adaptable, tailored to individual or team needs. These environments could blend physical comfort with virtual interfaces, offering personalisation beyond anything we can currently imagine. The idea is to move past the static office setup and towards dynamic environments that can morph into whatever the task demands—a brainstorming lounge one moment, a private concentration zone the next.

Innovations in data security and privacy are also crucial, given the distributed nature of remote work. As work continues to shift online, ensuring the integrity and confidentiality of data across networks is essential. Predictive analytics and blockchain technology could play pivotal roles here, preemptively identifying threats and ensuring secure transactions without the need for centralised control. The development of decentralised networks may offer unprecedented levels of security and trust, empowering both organisations and remote workers with peace of mind in their online interactions.

With these technological advancements comes the potential for enhanced inclusivity. Voice-activated and adaptive technologies can support diverse needs, enabling those with disabilities to participate fully in remote work settings. New innovations got the opportunity to break down longstanding barriers, thus promoting diversity and equity in ways previously thought impossible. Organisations should champion these technologies, not only as a matter of social responsibility but as an enrichment strategy that taps into the full spectrum of human potential.

Moreover, the evolution of remote work could instigate shifts in educational paradigms. With resources being redirected towards virtual learning environments, employees could engage in continuous upskilling seamlessly. This mode of education favours flexibility, allowing individuals to acquire new skills in line with emerging demands while integrating learning within daily work life. Innovations in this area might not only reshape how we think about education but also how we define professional growth and success.

However, it's not solely about technology. The human aspects of work—connection, motivation, and well-being—are equally critical areas of innovation. New strategies that focus on fostering community and engagement within remote environments will be pivotal. Virtual reality social spaces, gamified team-building activities, and AI-driven insights into team dynamics are just the beginning. These innovations could redefine how we connect with colleagues and nurture company culture even when physically apart.

As legislative landscapes adapt to these changes, we're likely to see new policies that accommodate non-traditional work setups better. Governments, along with organisations, will need policies that support a remote way of life, from tax regulations to international work visas. As the lines between local and global workforces blur, adaptations in policy will help maintain equity and ensure that remote work remains a viable and attractive option for employees and employers alike.

Finally, the environmental implications of a predominantly remote workforce must pervade our innovations. With less commuting, the promise of reduced carbon emissions is significant. However, this shift also requires eco-friendly solutions for increased home energy use, virtual data transfer, and equipment disposal. Sustainable innovation will be instrumental in ensuring that the environmental benefits of remote work are realised and maintained.

As we anticipate these innovations on the horizon, the opportunity to redefine remote work as a more efficient, inclusive, and sustainable model presents itself. Navigating this new frontier won't be without its challenges, but the rewards for those who embrace change could be transformative, offering a future of work that is adaptable, empowered, and profoundly human.

Chapter 25:
Personal Growth and Autonomy

In the evolving landscape of remote work, personal growth and autonomy emerge as cornerstones of professional fulfilment and success. As traditional frameworks dissolve, individuals are increasingly tasked with navigating their own paths, embracing the freedom to redefine personal and professional boundaries. This autonomy, while liberating, calls for a deep introspection and commitment to self-motivation; it prompts us to assess our skills, adapt to novel challenges, and cultivate an entrepreneurial mindset within our roles. Precisely because remote work demands that we become our own managers, setting goals and assessing progress takes on a newfound importance. In this environment, professionals must harness innovative strategies to remain inspired and driven, capitalising on opportunities for self-directed learning and skill acquisition. By doing so, they not only fuel their career trajectories but also contribute to a more dynamic and resilient workforce. The journey towards autonomy is not without its hurdles, yet it ultimately empowers us to take ownership of our development, offering the opportunity for unprecedented personal and professional enlightenment.

Embracing Independence

In the evolving landscape of remote work, embracing independence has emerged as a cornerstone of personal growth and autonomy. It's about seizing the opportunity to redefine how we manage our work and personal lives, all while pushing the boundaries of traditional office dynamics. Professionals are finding that independence offers a unique blend of freedom and responsibility, where self-direction fuels creativity and innovation. With the liberty to shape one's schedule comes the obligation to cultivate self-discipline and accountability. It's a transformative journey that compels individuals to harness their intrinsic motivation, fostering skills that transcend professional boundaries and ripple into personal development. This shift in mindset allows professionals to become not only adept problem solvers but also pioneers of their own destiny, crafting a work-life paradigm that champions flexibility and self-reliance. In this brave new world, independence isn't just a skill—it's a catalyst for unprecedented personal and professional evolution.

Strategies for Self-Motivation form the bedrock of embracing independence in a remote work setting. Embracing independence isn't just about relishing the freedoms remote work can offer. It's about seizing the reins of your professional journey and recognising that self-motivation is your fuel. Without the typical office environment, you're required to construct a framework that aligns with your personal and professional goals. In this digital arena, self-motivation isn't a mere survival strategy—it's a catalyst for thriving.

One cannot underestimate the significance of setting clear and concrete goals. Goals act as guiding stars in the vast night sky of professional independence. Start with ambitious, yet attainable objectives that align with your long-term career aspirations. Splitting these into smaller, manageable tasks can create a sense of achievement as each milestone is reached. Specific goals provide clarity, ensuring you're constantly moving forward rather than drifting aimlessly in the sea of remote work. Remember, successful self-motivation is as much about the journey as it is about crossing the finish line.

Another effective strategy involves building a routine that supports your productivity and wellbeing. A flexible schedule is one of remote work's most attractive propositions, but it is essential to craft a routine that helps you maintain focus. Begin your day by setting clear intentions: what are the key tasks you aim to complete today? This proactive approach carves out mental space for you to operate within, giving your brain a framework within which creativity and productivity can flourish.

Rituals can also play an essential role in enhancing self-motivation. Whether it's starting your day with a few moments of mindfulness, a morning workout, or simply a warm cup of tea, these rituals can become anchors that ground you amidst the fluidity of remote work. They create a rhythm that sustains motivation and lets you start each day on a positive note.

Moreover, embracing independence means welcoming a growth mindset where failures are viewed as learning opportunities rather than insurmountable obstacles. This mindset encourages you to experiment, learn, and improve constantly. Inject a dose of curiosity, and you might find yourself eagerly awaiting challenges, ready to dissect them and adapt. In the absence of physical workplace mentors, become your own guide by harnessing resources like webinars, online courses, and virtual communities related to your field.

For professionals navigating remote work, maintaining connections with colleagues even in a virtual space can be a powerful motivational tool. Building and nurturing relationships with like-minded individuals fosters a sense of belonging. This network can offer support and inspiration, driving you to achieve the highest standards of performance. Lean into virtual coffee breaks or professional group chats; these interactions can serve as a reminder that though we may work alone, we are part of a greater community.

Another facet of self-motivation lies in creating a physical environment that encourages productivity. Design a workspace that reflects your professional and personal style. This space should be both functional and inspirational. A well-organised desk, ergonomic seating, and natural light can significantly affect your attitude towards work. Even a small plant or favourite photograph can inject positivity, stimulating your motivation levels.

While focusing on enhancing motivation, it's crucial to acknowledge and celebrate achievements—be they big or small. Recognition isn't just a tool for managers; when applied personally, it can rejuvenate your drive and make the journey fulfilling. Celebrate completion of tasks and reflect on how far you've come in embracing this new independence.

Engaging in activities outside work that nourish your soul is equally vital. Hobbies and passions unrelated to your profession can be energising, providing fresh perspectives that reinvigorate your work enthusiasm. Balance, after all, is a silent partner in motivation, offering a rounded approach to embracing independence.

In this era where traditional work narratives are constantly being rewritten, self-motivation serves as the ink. As we scrutinise and redefine what professional success means, let the strategies mentioned be your toolkit. The shift to remote work empowers us individually, urging us to venture forth and plot our paths. In doing so, we not only enhance personal growth and autonomy but also contribute to a rapidly evolving work landscape.

Conclusion

A s we draw together all the insights, strategies, and reflections shared throughout this journey into the realm of remote work, a vivid tapestry emerges. This transformation isn't merely a paradigm shift in the work model; it's emblematic of a broader evolution in how societies, companies, and individuals operate and thrive. Each chapter of this book has unravelled a facet of remote work that collectively points towards a future brimming with potential and challenges. Together, they scaffold a reality where remote work is not just supplementary but central to the way we envision organisational life.

Acknowledging the pivotal role of technology, it becomes clear that the digital tools and innovations that catalysed remote work have also democratised opportunity and access, transcending geographic and temporal barriers. The symbiotic relationship between technology and remote work calls for ongoing adaptation, both from individuals who must continuously hone their skills and from organisations that need to flexibly integrate emerging tools. This technological underpinning ensures that remote work can be as effective, if not more so, than traditional office setups. It's not merely about adopting tools but about embedding a mindset of continuous innovation and improvement.

The intricate balance between our professional and personal lives, long a subject of concern and debate, takes on new dimensions in the remote landscape. Rather than simply juggling responsibilities, remote work demands an integration of phases, facilitating a lifestyle where personal and professional growth can coexist harmoniously. Organisations that support this balance by fostering environments respectful of their workforce's diverse needs and timezones, often emerge not only successful but also deeply resilient.

The psychological aspects of remote work underline the necessity for robust mental health support systems. As our kitchens, living rooms, or bedrooms double up into office spaces, the lines between 'work time' and 'personal time' may blur, potentially impacting wellbeing. Successful remote work strategies incorporate mechanisms for emotional resilience and endorse the importance of mental health. Building a supportive community, even virtually, is paramount to sustaining morale and productivity.

Communication, now more than ever, stands at the heart of remote work success. It's more than maintaining frequent touchpoints—it's about fostering genuine connections. Effective communication channels are the threads that bind remote teams; they must be both comprehensive and reflexive, adapting to the nuanced needs of diverse workforces. Leaders have a special role to play in nurturing these communication pathways, ensuring authenticity and inclusivity throughout.

Leadership needs rethinking in the remote context. It must shift from oversight to empowerment, from micromanaging to trust-building. Leadership effectiveness now hinges on the ability to inspire and motivate without physical presence. Here lies the blueprint for future leaders—those who value autonomy, encourage innovation, and build teams that thrive on mutual respect and goal alignment.

Indeed, remote work has redefined corporate culture, offering an array of opportunities for companies to innovate in how culture is built and maintained. It challenges organisations to rethink how they can create a sense of belonging and shared purpose from afar. Ways to engage employees, celebrate successes, and communicate values have evolved, becoming more intentional and inclusive.

The spectre of declining productivity often looms over remote work, yet, paradoxically, it offers a unique canvas for excellence. By focusing on output rather than hours, individuals and teams can redefine productivity on their own terms. The strategies discussed throughout offer a pathway to harness both time and energy more effectively, fostering not just the achievement of goals but exceeding them in novel ways.

From the perspective of career development, remote work can be incredibly empowering, providing unique opportunities for growth, education, and skill enhancement that are not constrained by physical borders. This opens new avenues for companies and workers alike, enabling careers to flourish irrespective of geographical constraints and inviting a culturally richer fabric of workforces.

The avenues for inclusion and diversity also expand vastly. The changes in hiring practices, facilitated by remote work, allow businesses to tap into a wider talent pool and benefit from the array of perspectives this brings. It urges organisations to build environments that respect differences and leverage them as strengths—an imperative for modern business.

Remote work also introduces a shift in the landscape of real estate, public transport, and infrastructure planning. As fewer people commute, there's potential for profound environmental benefits alongside the necessity for innovative utilisation of traditional office spaces. This evolution prompts a reimagining of urban design and residential preferences, steering towards sustainability and efficiency.

Finally, the essence of personal growth and autonomy finds a new champion in remote work. Workers are urged to take ownership of their careers and lifestyle, challenging them to rediscover motivation in their daily routines and gifting them the freedom to design their work lives around personal strengths and preferences.

In conclusion, remote work is not merely a temporary adjustment. It's a door to a re-envisioned future of work that is more flexible, inclusive, and resilient. While challenges remain, the opportunities to innovate, connect, and grow are boundless. As professionals and leaders, the charge is to embrace this transformation—we are the architects of this new world of work. By fostering a mindset attuned to adaptability, creativity, and empathy, we can craft reality where the benefits of remote work are harnessed for the betterment of all, paving a way for progress that is inclusive and sustainable.

Appendix A:
Appendix

The journey through the landscape of remote work has been both profound and transformative. As we pause to encapsulate the myriad insights and discussions detailed in the preceding chapters, this appendix serves as a reflective tapestry, weaving together the threads of knowledge, challenges, and opportunities encountered along the way.

Remote work has redefined our professional boundaries, reshaping not just how we work but where and when we choose to engage with our tasks. As we've analysed the integration of technology, the reshaping of corporate culture, and the facets of health and wellbeing, it's evident that this seismic shift offers a canvas ripe for innovation and personal growth.

Throughout this book, we've explored the complex interplay of elements that underpin successful remote work environments. It's become clear that agility and adaptability are not mere buzzwords but core tenets necessary for thriving in this new era. The art of leadership, communication, and productivity has evolved, demanding a fresh perspective and novel approaches to foster engagement and efficiency.

Moreover, the ripple effects of remote work stretch beyond individual sectors, influencing global economies, educational paradigms, and even the urban landscapes we inhabit. Understanding these impacts equips us to navigate and shape the future thoughtfully. This appendix is not a conclusion but a springboard, encouraging readers to continue exploring the nuances of remote work and its broad implications across various domains.

As we stand at the precipice of this transformation, the challenge is not merely to adapt but to harness change proactively. We urge HR leaders, corporate managers, and professionals alike to remain curious and committed to fostering environments that are not just reactive but creatively responsive to the demands of remote work.

The lessons drawn here serve as a guide for embracing the autonomy, inclusivity, and sustainability that the remote work ethos champions. Let's venture forward, equipped with the insights garnered, poised to not only meet the demands of this evolving landscape but to flourish within it.

www.ingramcontent.com/pod-product-compliance
Lightning Source LLC
Chambersburg PA
CBHW051429280526
45785CB00003B/1213